BEATING ABOUT THE BUSH
RANDOM MEMOIRS OF AN EX-BRIT

BARRY COTTON

Copyright © 2012 by Barry Cotton
December 2012

ISBN
978-1-4602-0312-5 (Paperback)

All rights reserved.

No part of this publication may be reproduced in any form, or by any means, electronic or mechanical, including photocopying, recording, or any information browsing, storage, or retrieval system, without permission in writing from the publisher.

Produced by:

FriesenPress
Suite 300 – 852 Fort Street
Victoria, BC, Canada V8W 1H8

www.friesenpress.com

Distributed to the trade by The Ingram Book Company

FOR MY FAMILY

Contents.

INTRODUCTION	1
1. EARLY DAYS	5
2. SCHOOL DAYS	23
3. WAR-TIME.	45
4. BRITISH COLUMBIA	93
5. DOWN UNDER.	203
6. HOME AGAIN	229

INTRODUCTION.

"Autobiographies ought to begin with Chapter two"

– Ellery Sedgwick (1872–1960)

Most of these episodes were written about shortly after they occurred; with the exception of "Early days", which - memory often being keener when it concerns childhood - is just as authentic, although described some sixty years later.

There was also the odd person who said: "What makes you think anyone will want to read your memoirs?" and this person had a point; although the answer can be complex. Events of the past will usually hold a readers' interest if his or her own memories reach back to the same era. Although my generation is nearly gone, it will not become extinct provided that there are writings which bring back familiar events of the past to succeeding generations. In my ninety four years I have seen a lot of country, and if the episodes herein give a glimpse of a past generation, its foibles, its achievements, then the book will be found interesting.

If I were in the public eye, there would be no lack of potential readers. In fact, I would hardly need to labor over producing this chronicle on my own account - I could employ a ghost-writer. Yet affluent or otherwise, I would not even think of doing so, for the simple reason that I have thoroughly enjoyed the writing of it.

I have read many biographies in my time, and found them, on the whole, more interesting than fiction. Life stories can be inspiring, exciting, romantic, or shocking, but nearly always interesting. Mine may not aspire to any of the former attributes, but I am hopeful that it may be found entertaining.

H.BARRY COTTON

My grandparents James Liggins Cotton and Annie Cotton, with my father, c - 1888.

The church at Clifton Hampden, Berkshire, where J.Liggins Cotton was vicar. A picturesque English village.

FAMILY ARCHIVES.

My Uncle Hugh and cousins Tom & Jim at Woodmancote, Bucks, c - 1928.
The rabbits were not amused.

Cotton Grandchildren 1936
Standing -
Tom, Barry, Eleanor.
Sitting -
Jim, Diana, Gran (Annie) -
Cotton, Anne, Geoffrey.

Gran Clampitt, Uncle Leslie,
Aunt Do, and Audrey (seen laughing!).

My mother in 1918. She was a beauty as I was well aware.

The next generation.
Wendy, my mother, myself, David and Kitty 1956.

3

Canadian cousins - Kitty, Alena, Barry and Bruce 1949.

Canadian Aunts. Minnie & Daisy c -1930

Kitty & I 1950.

MORE FAMILY ARCHIVES.

Australian cousins. Eleanor & Paddy.

Welsh cousins - a boatload: Gwenfrewi, Mair, Gorwen, Non, & Ceinwen. 1956.

1. EARLY DAYS

BEGINNINGS

"What is the use of your pedigrees"

– Juvenal (AD 50–c. 130)

The non-event in which I played a leading role occurred on December 17th 1917, at a house called "Landore", on Stoney Hey Road, New Brighton, England. The Wirrall peninsula in Cheshire, has produced many famous persons, not a few infamous ones, and a great many more, such as myself, with no claim to distinction whatsoever. My early association with the Wirrall was quite short, and I spent my youth in a variety of other locations.

Although I was christened with the names "Hugh Barrington", the "Hugh" was never used, as I had an uncle on my father's side of the family who also went by that name, and obviously there could not be two "Hughs" in the group.

So I was labelled "Barrington" as well. I had a suspicion that this name was pulled out of the air, and pinned on me, because it sounded genteel; my mother, at eighty-eight years old, confirmed this when I asked her. It was not so surprising, as, if there was one thing which both sides of the family considered important (and they were worlds apart in other matters), it was social standing. I did not regret the name, however. When I was young, there were certain advantages to it as a form of address. When shortened to "Barry", it meant that all was well, and relationships were cordial; when pronounced in full - "Barrington!" - it usually boded ill for me, and warned of heavy weather in the offing.

My father's family came from the south of England. The matriarch - my grandmother - was the widow of a country clergyman, who died in the late 1800's, leaving her with four young children to bring up, and with limited means. She was high-principled and devout. She was also indomitable, and often domineering. To some - my younger cousins - she was quite literally a holy terror; to others - my mother, for instance, she was downright cruel and vindictive; to me she was affectionate and kind, and she looked after me for several years, when my parents had no home of their own. She was always a pillar of the Anglican church, and she introduced me to religion quite early in life.

I said my prayers (in true Christopher Robin style) under her supervision, and was regaled with stories of the New Testament regularly at bedtime. Details of the Crucifixion, however, did not appeal to me, and I sometimes reflect now that, virtuous as my grandmother might have been, imparting such horror stories could hardly be in the best interests of a small child. Fortunately there were other bed-time stories - many of which gave me glimpses of other lands, and other ways of life - such as Ernest Thompson Seton's timeless tales (who has now heard of them?) - Kipling's Jungle Books - and even Tom Sawyer. She was a keen and conscientious reader of bed-time stories, and I listened with pleasure. The affection which I received from her, and which I certainly returned, may have been due to the fact that I was the eldest grandchild, but it was very necessary to me.

Her family consisted of my father - the eldest son, tall and aloof, a daughter married to a tea-planter in Southern Madras, India, a second son - also a tea-planter, married to a charming New Zealand girl, and the youngest son, my namesake, an officer in the Royal Navy. All of them visited her home from time to time.

By the time I was born, the family were in a state which is often described as that of "genteel poverty". What they lacked in money, however, they made up for in the strength of conviction of their own superiority. At such a young age, as I was, a child accepts adults without question, but it is safe to say that, with two exceptions, I found them an unamiable lot. My mother was to say later that, if there were any blue blood in those autocratic veins, it was certainly tinged with ice.

On the other hand, my mother's family were not pretentious. They were aware of their social standing (who is not, in Merrie England?), but they were not - as the Cotton's and the Pharisees - arrogant in their beliefs that they were "not as other men are". Prior to the first World War, my maternal grandfather, an enterprising man, who attended the Bluecoat school in his youth, had started a wholesale meat and bacon importing business in Liverpool. Then, as now, Northern Ireland and Eire did not see eye to eye on many matters, trade in meat directly across the border being one of them. My grandfather found that, although the two countries would not trade directly with each other, they had no objection to trading through a middle man. He set up a very lucrative business, buying from Eire and selling to Ulster, and by the end of the first war was comfortably well-off. He had a large house in New Brighton, and on important occasions would go out in a hired car with chauffeur. I am told that in the pre-first-war days, he took his wife to Egypt on two occasions, a trip that only the rich could afford.

Eamon de Valera, when he became Prime Minister of Ireland in 1932, put a stop to the trade practice on which my grandfather's business was founded, and after carrying on at a loss for two years, he eventually closed up and retired. In his heyday, he was a forthright and decisive man, and ruled his family as most Victorian parents did - with a heavy hand. He died in the 1940's, outlived by his widow, my grandmother, who attained the age of 96 before she passed away. The aforesaid heavy hand was responsible for his eldest son - my uncle Allan - leaving England at age 16. He did not get along with his father at all, and means were found for him to emigrate to Canada, where another branch of the Clampitt family had already settled in Saskatchewan, and would take him as an apprentice. He soon moved to British Columbia, and it was there - in 1946 - that I met him on my own arrival in lotusland.

The other Clampitt children consisted of a younger brother, Leslie, and two sisters, Gladwys and Ethel; Ethel, the youngest, was my mother.

Whereas the Cotton family rejoiced in the possession of a family tree, extending back to the year 1400, I have not been able to learn much about my maternal grandfather's antecedents. There are rumours of a Welsh connection, the forbears of my grandmother. My grandmother also claimed descent from one General Outram, a famous man whose orders regarding the greasing of cartridges with pig-fat are said to have been one of the causes of the Indian Mutiny. Ancestors, I feel, should be accorded their proper due, and this one is a real prize-winner. Although I couldn't possibly go as far as Peter Ustinov, who in his autobiography starts by naming and describing all sixteen of his great-great-grandparents - a truly herculean feat.

My father had been educated at Tonbridge School, then Cambridge University, where he received a B.A. After his death I remember looking through page after page of school photos, taken at Tonbridge (the 3rd XV rugby team - the 2nd XI cricket team - the 1st XV rugby team etc. etc), in which groups of deadly serious young men gazed steadfastly at the camera, while the one in the centre held a ball (or a cricket bat, or maybe a trophy), and marvelling that not one of these young men had the ghost of a smile on his face. Presumably , if one of them had the temerity to grin, the photo would have been discarded, and a new one taken. I do not remember my father smiling much in later life, either, but maybe he didn't have much to smile about. He served in the first World War as adjutant of the 3rd Cheshire Regiment, from which activity he acquired a chronic case of varicose veins. Whilst in uniform, he married my mother. After the war, he attempted to put his education to good use by starting a co-educational school in a little village, Bridge-of-Weir, in Scotland. It was known as Ranfurly Castle School, and consisted of a colossal stone building that dominated the railway station in that small place, which now is a thriving suburb of Glasgow.

Here, my first childhood memories began, and until the age of five and a half, when the school went bankrupt, I enjoyed a fairly stable existence. Life, of course, had its ups and downs. Because of he social ideas of the upper-class English at that time, I was not looked after by my mother, but relegated to a nursery, on the upper floor, where a nanny presided. Every evening before supper, I was brought downstairs to be with my mother for an hour, after which - back to the nursery (the *proper* place for children). Mothers in this society were not expected to do such things as changing diapers - that was for the hired help; and since the nannies were supervised by my paternal grandmother, they came and departed fairly regularly. My mother, I did know and recognize, even though I only spent a limited time with her. Of the succession of nannies, I only really remember the last one, of whom I was very fond, and who took me to stay with her "folks" in the Isle of Man when the school was closing down.

Recollections before the age of five can be pretty hazy. But impressions do remain, and there are some which always have been quite vivid. Why I should remember being duly set down on a pot every day after breakfast, I do not know, but I even remember the look of disgust on my nannie's face, when, elated with success, I triumphantly proffered the pot for her inspection. If this was toilet-training, then memories can indeed go a long way back. I remember long country walks, a scooter, and a kitten (which was finally taken away, probably for its own protection). But other children do not figure in the recollection, mainly because they weren't there. I was the only occupant of that nursery. On one occasion I wandered into the bathroom where Mary Dudley (the daughter of my father's partner, and a few months younger than I) was being bathed. I was quickly shooed out of the door before I could ask any questions; however, I had seen whatever it was that the adults didn't want me to see, and I forthwith resolved that one day I would find out more about this curious phenomenon (although it was not till several years later that I was able to satisfy my curiosity). I remember learning the Highland Fling - my mother was the teacher, who would have taught dancing regularly in the school, had not my father vetoed the idea. Hemmed in by both my father and paternal grandmother, my mother was much oppressed.

At five years old, I was taken out of the nursery for short periods, and attended a kindergarten class in the school. We learned our ABC's here, and did all the usual kindergarten activities. We all recited a poem at the

end of term - to the assembled parents. This I can so vividly remember, that I can even quote the poem. It went as follows:-
"The life of a snail is a fight against odds,
Though fought without fever or flummox;
You see, he is one of those gastropods
Which have to proceed on their stomachs".

The fact that none of us had the slightest idea of the meaning of any of these long words - or even of the poem itself - did not worry us at all. We stood up, held forth and duly received our applause. One of the boys had a broad Glaswegian accent; *his* poem went:-
"The wurrms, the wurrms, the wriggling wurrms! . . . . . "

I even wondered why all the parents laughed so much; but at that age, adult behaviour is seldom questioned, but simply accepted.

The teacher of that class was a Miss Brown, under whose supervision we all sat at small desks. It was Miss Brown's task to impart to us the first rule of school i.e. that the pupils must pay attention when the teacher speaks. I know that I was a fairly unruly pupil, and I'm afraid that I really tried Miss Brown's patience. She would often make little jokes to the class, as she drew pictures on the blackboard, but whereas most of the other pupils simply tittered dutifully at them, I would usually open my mouth wide, and give vent to a loud "Har! Har! Har!". One day Miss Brown's patience collapsed, and she marched me in to see my father, the headmaster, in his office.

The resulting chastisement that I received - and obviously deserved - did much to cut me down to size. My father, after trotting out the hoary old adage about it "hurting me more than it hurts you", bent me over a chair, and administered four light pats with a ruler. I was most upset, and was delivered back to the nursery for the rest of the day. Next day in class, I was suitably contrite. Whenever there was cause for laughter, and I felt the urge to break forth with a guffaw, the glint in Miss Brown's eye was enough to stop me after only one, or at the most two "Har's".

It must have been in 1922 that the school closed its doors. The capital had been provided by my father, his partner, and various family members. All lost their money, and my father was literally broke, and unable to find a job for two years. My nanny suggested that she take me to the Isle of Man, which was her home, and look after me until the question of where we would live had been settled. The suggestion was adopted thankfully by my parents, so off I went with Brenda, my Nanny (with a capital N now), by way of the ferry boat from Liverpool. It was rough passage, and inevitably I threw up all over the cabin floor.

The Banks family had a house in Ramsay, and several children. The nearest to my age was Nigel, maybe a year older. There were many other neighbourhood kids, and the boys tended to go around together in groups. I was the youngest in the group, and "tagged along", learning all kinds of new things. We raided apple orchards, bowled hoops down the main street, collected discarded cigarette butts which we packed into a home-made pipe and smoked (it took several days to collect the butts, but the smoking only took about two minutes each before we turned green, and we didn't repeat the experiment). The escapades in which we indulged were pretty harmless; however, it was impressed on me by the older boys that on no account must I tell an adult what we were up to, as most of the time we were engaged in doing things which we had been expressly forbidden to do. There were family picnics on week-ends, and when the T.T. Motor cycle races took place we sat on rugs beside the track. There was no radio commentator to describe the action in those days, but the older boys knew every rider, his history and every detail of his machine. I remember that one of the adults even organized the boys into a time-trial race of our own, complete with competitor's numbers and home-made crash helmets (we ran on foot, but imagination supplied other details such as handle bars, gear levers, and we made very realistic engine noises).

Mrs. Banks was a very good mother, and I don't remember any unpleasantness while I was in her house. On one occasion, however, she did have cause for vexation.

The popular songs of the day, which we sang, hummed, whistled, and often parodied when we didn't know the words, were:- "Yes, we have no bananas", "Tea for two", "Bye, bye blackbird" - (the *first* time around), and a song called: "Horsey keep your tail up". This last one never did

13

attain such popularity as the others I have mentioned, but we sang it a lot.

The first and second lines of this song went:

"*Horsey, keep your tail up, your tail up, your tail up; To keep the sun out of my eyes'.*

It happened at dinner one day, that one of the adults - while bringing in a dish from the kitchen - sang the first line; I, in all innocence, carried on with a parody of the second line, which we kids (when out of ear-shot) had been wont to sing:-

"*. . . and I'll stick a nail up!*" (It *did* rhyme well, after all).

After the first moment of shocked silence (there was no doubt that I was out of order, as young Nigel was shaking his fist at me covertly) - I was asked to repeat the phrase; after which a hail of retribution descended, not on my head, but on young Banks for having taught me such an improper expression. Needless to say, I was not popular with the other boys for a few days afterwards.

Towards the end of the summer, my mother came to fetch me. I remember being sorry to leave Ramsay, and I never saw Brenda again. But it had been decided that I was to live at Gran Cotton's house at Tongham, Surrey, until my father had a job, and could provide a home for us. From then on, my destiny was in the hands of the paternal side of the family, and my mother had little to say in the matter.

TONGHAM

"Ones own relatives are always the worst"

– Norwegian proverb.

The house in Tongham was known as the Manor House. It was situate in two acres of grounds, a large part of which was flower garden. There was also a kitchen garden, chicken run and several old sheds (former stables) at the side of the house. The front garden had flower beds, with colorful dahlias. My grandmother was a keen gardener, although a "man" was employed to do such things as digging. There was also a cook, and a maid. The state of being poverty-stricken never seemed to extend so far as having the family do their own household chores.

Like all old English houses, this one was cold and damp. There was a fireplace in every room, with a coal-scuttle beside it. The luxury of electricity was yet to come at that time, but we had gas lamps with mantles as fixtures in each room. I used to take a candle upstairs to bed, and the window was usually open during the night in all seasons (sixty-five years later this still seems to be the practice on the part of many English people, who don't feel healthy unless they wake up in the morning with cold noses). I used to have a small kerosene lamp by the bedside - a night light - when aged six, to dispel any fear of the dark; it was an old and creaky house, and I still can remember the apprehension I felt at night, when the wind blew in the trees outside, and I wended my way upstairs with a flickering candle. However, once abed, someone would usually come upstairs to say goodnight, and I slept reassured.

This house contained my grandparent, father, mother, various uncles and aunts periodically, and myself. Relations amongst the adults were not too harmonious, although I was too young to take notice of such a state of affairs.

My mother had married my father in 1915, after an unhappy previous engagement. It was a grand military wedding, the bride in white, with a long train, and archway of crossed swords and so on. The bride, however, was no society debutante, but simply an unsophisticated girl of less than twenty. There is no doubt but that she was a beauty; the photos attest to that; in fact she maintained her attractiveness up to, and including the days of her old age. Like most young ladies of that era she had learned the social graces - to play the piano, sew, cook and do household chores. She was, I am sure, a conscientious wife to my father.

But two more unlike personalities could hardly be imagined. He was tall, reserved and studious; she was gay, frivolous, fond of people and parties, and - throughout her life - fascinated by pretty clothes, jewelry and fine furnishings. At various times in her life she was able to gratify her tastes in these respects, but in the early days of marriage to my father there was little money to spare, so she had to be satisfied with what was (to her) a rather dull existence. Although some women were able in those days to have careers, I don't think she would have even considered it. Apart from a period of a few months during the Second World War, when, already divorced twice, she was a civilian chauffeur for the de Haviland Aircraft Company, she never earned money by actually working, but managed to make her way by renting rooms, buying small houses, having them renovated then selling at a profit. At this activity she became very expert. Finally, in her late fifties she inherited part of her parent's estate.

But at the time of which I am writing, she was inexperienced and naive, and Gran Cotton's roof was not the place under which a carefree young girl could live unoppressed. Oppressed she was, from the moment she came; looked down upon, snubbed and ordered about; until - her husband being as he was, dominated by his mother, and incapable of standing up for her - she finally left the house and went home to her own parents, in spite of his protestations.

I think I must have gone with her at first, as I do have very early memories of Christmas at New Brighton. It was a unique Christmas in that - even at the tender age of seven - I got to drink a glass of wine at dinner. (At Gran Cotton's house, alcoholic beverages of any kind were strictly *verboten*). Also, I met my cousin Maureen, who was four, and we went roller-skating together on the sidewalk on Stoney Hey Road. *And*, by the process of staying awake long enough, I found that there wasn't really a Santa Claus, only my uncle, without even a red suit for the occasion.

There was a large attic in the house, full of the most interesting things - a rocking horse, and many small Egyptian souvenirs; also a collection of birds eggs, all catalogued in trays, which, I later found out, belonged to Uncle Allan. Fortunately for us, he was safely in Canada, as we left no egg unsmashed. Maureen was the daughter of my mother's elder sister. Both of us enjoyed each other's company whenever we met, then and in later years. There was also a colored maid-of-all-work, Sarah, a very kindly soul, and much of my time was spent in company with her and Maureen. I remember nothing but kindness from all the members of the Clampitt family, and often stayed with them in later years, and when on holiday from school. It always seemed to me strange (and still does), that they spent much of their time feuding with each other, and in old age hardly communicated at all.

That Christmas was only an interlude. When it was over I returned to Tongham, where a new treat was in store for me. I was to go to school.

Barry Cotton

DOWNE HOUSE.

"I confess freely to you, I could never look long upon a monkey, without very mortifying reflections"

– William Congreve (1670–1729)

There was no doubt that I posed a problem for the Cotton family. Since my mother had gone to live at her parents' house, and my father was out looking for a job, someone else would have to look after me. My grandmother was quite willing to do this, but there was also the question of "civilizing" me. When I first arrived at that house, I used various strange expressions picked up in Scotland, such as "Havers!", and "Och, awa'!"

The school chosen had one claim to being unique. It was situate in the house that had belonged to Charles Darwin, in the village of Downe, Kent. It was called Downe House school. This was the house in which "The Origin of the Species . . . " had been written, and the room which had been Darwin's study was kept apart, as a shrine. I remember hovering in the background when some visitors were being shown around, and the door opened while the the headmistress said: "Yes, this was his study . . . . ". Why this should stay in my mind I do not know, but it is probable that we, the younger pupils, had been told about Darwin, and that he was a famous scientist. The house was a large one, and able to accommodate the fifteen or so pupils and staff. The downstairs room had been converted to classrooms, dining and day rooms; upstairs were dormitories. As there was only a small staff of servants, many of the household chores were done by the pupils - bed-making, sweeping and dusting. Outside were extensive grounds, including a large lawn with a mulberry tree, a kitchen garden, shrubbery and a tennis court.

The fifteen pupils ranged from the youngest, Valerie, who was a little younger than I, to the eldest, Beryl, who was sixteen. Beryl was called "Bee" by all and sundry; she was the "head girl", and I adored her. There were only two other boys, one called David, the other called David Pugh , and they were both my age. All the other pupils were girls, many of whose names I can even now remember. As far as teachers are concerned, however, my memory is hazy, but I do remember the headmistress. Mrs Ram was a rather masculine female with a long nose. (I have mistrusted people with long noses ever since those days). She usually wore tweeds, and I do not ever remember seeing her smile.

Lessons took place in the mornings only, the afternoons being devoted to other pursuits, such as walks in the country, or to the village (where we walked in pairs, in true crocodile formation). Now and again there was elementary tennis, and always there were more walks in the country. On one exceptional occasion, we went to the movies (silent of course), and saw a cowboy film, featuring Tom Mix - a glimpse of a different world. We were also introduced to the Girl Guide movement (for the boys it was Wolf Cubs), where a Joyce-Grenfell-like lady - the "Brown Owl" - taught us about toadstools, and faeries, and how to salute with two fingers, how to sing "Land of Hope and Glory", and all about reef knots and granny knots (with which I still have trouble).

In class we did copy-book writing, spelling, elementary geography and history, and we chanted the mathematical tables out loud (which seemed to be the standard way of learning them in those days:- "One and one are TWO! Two and two are FOUR! and so on). It was found at one of these sessions that I could not read the blackboard; a visit to the oculist followed, and soon my face was adorned with a pair of spectacles - labelled as "gig-lamps" by the other pupils. The trouble which I got into each time I broke them did not seem to lessen the frequency with which I did so. I think I was an extremely unruly pupil. and remember many times being "stood in the corner" - but then all of us suffered that punishment frequently. At least there was no "dunce's cap" in use in that school.

On Sundays we had Church, all dressed in best school uniforms; and we all had to write home. My first letter, to my grandmother, was to the effect that I was having a perfectly foul time at this place, and I didn't like it, and would she please come and take me away. The letter was promptly returned to me by the teacher, who said that my family should not be given the impression that I was unhappy, as it would displease them, and, in any case I was *really* enjoying myself, so I should write about that instead. So the letters never did say much more than:- "I hope you are well, we had treacle pudding for dinner".

During the school term, it was possible for parents to visit their offsprings, and take them out - for tea, maybe - or for a walk. I remember two such visits; one from my father. We went for a walk, the two of us, and while I chattered incessantly as a seven-year-old would, it was quite noticeable that he himself hardly said a word. Looking back, I think it highly likely that his mother pressured him into coming to see his son at school; even to me, he seemed somewhat embarrassed, as though he didn't know quite what to say; and that was the first, and last visit I had from him at any school that I attended, all being boarding schools.

On another occasion, my mother came to visit, and she brought a small cake. In keeping with the proper ritual for pupils who received goodies from home, I went around the dining room tables, offering everyone a piece of it, and it was so politely refused by them all that I felt quite miffed, until I finally came to Valerie. "Oh, yes please!" said Valerie, just as politely, so I cut her a large wedge, and between us we ate half the cake; which was just as well, as the cake mysteriously disappeared after that meal, and I have no doubt that the kitchen staff finished it off. My plaintive enquiries as to its fate were completely ignored.

I remember Valerie with affection. She it was, one afternoon in the shrubbery, who demonstrated to David Pugh and I the difference between little boys and little girls. (Probably the most useful learning experience I gained in that school). Mrs Ram, however, soon got word of these goings-on, and, after verbally castigating Valerie (she was a girl), administered the proper punishment for boys to me, with a ruler on the bare bottom. My howls were more due to indignation than pain. David Pugh, however, was dealt with differently by cunning Mrs Ram; and it became fairly obvious, after a return visit by the three of us to the shrubbery (to refresh our memories), that David Pugh had become a fink. No sooner had we entered the house than retribution fell on the two of us quite violently, and both our backsides were pummelled without mercy, while David Pugh received a kindly pat on the head for his part in the affair.

I found that going to sleep in a room with other kids was quite difficult. One does not go to sleep simply because one is told; and if there were someone to talk to, I talked. The solution was to put me in an alcove so that I was out of earshot. *That* was a mistake, and I proceeded to prove to all and sundry that I *was* within earshot. The senior girl, "Bee", came round one night, "tucked me in", and told me to go to sleep - I would do anything for her - and I did so. So every night regularly, my determination to be "tucked in" by "Bee" asserted itself (in spite of the exhortations, which I could hear, to "Bee" that she shouldn't pander to me), and I simply raised the roof until she came.

I like to remember this as one of the few battles (child versus adult) in my youth, that I won. But the next battle I was to lose. My father, as a preliminary to sending me to a new school, for boys only, decided to teach me the multiplication table at home; a laborious procedure, enforced without humour, which lasted until I was most unwillingly able to recite as far as twelve times twelve.

2. SCHOOL DAYS

SUSSEX-BY-THE-SEA.

"Of all the animals, the boy is the most unmanageable"

– Plato (428–348 BC).

There was a clothing list, which my grandmother studied diligently. It consisted of such items as:- "Drawers, winter, 4; drawers, summer, 4; handkerchiefs 12; pyjamas 3; and included a "tuck-box", a very important piece of furniture, the only place - it turned out - where one could keep anything personal or private at all, such as letters from home, which one didn't want other boys to read. At this school our normal attire was to be: black shoes - long stockings with the school colors - grey shorts - white shirts with that sadistic invention on top, the Eton collar, a stiffly starched horror which folded over the jacket, and held a tie with the school colors. I *think* we wore jerseys, too - but , whatever it was, we all wore the same, and the colors were purple and white. The outfit was topped off by a peaked cap (again purple) with a single button on the crown.

We wore this uniform until our last year, at aged thirteen - knobby knees and all - except on Sundays, when we wore a Cheviot (dark) tweed suit, with long pants, and a black tie with the Eton collar. Perhaps it was unseemly to display one's knees on that day of the week.

I was eight years old when I joined that school. There were two other new boys - one called Cedric, who, as soon as we were out of adult-earshot, told me a dirty limerick (all about an old man of Jamaica . . . . . and the embarrassment that ensued due to his insistent use of brown paper). I found it most amusing, even if it didn't rhyme very well, and lost no time in telling it to the other boy, Harley. Harley, however, was quite shocked. . . .

The school consisted of several adjoining buildings in a terrace overlooking the sea-front at Brighton. Mr. Arnold and Mr. Gausson were the two headmasters. Mr. Gausson was a likable man, a good teacher, and to me - even though I had the misfortune to be caned by him several times in my last year - a fair one. I deserved all I got. Mr. Arnold, with whom I had no contact in respect of teaching, was a portly and fairly handsome man. One of the advices which he gave his youngest class, on the subject of personal hygiene, was that each mouthful of food should be chewed twenty-four times at least. This was an interesting theory to the new boys of our term, and we would often watch at meal-times - the staff sat at a separate table - and silently count while he masticated. We found, however, that he seldom exceeded fourteen times. Naturally we didn't have the temerity to point this out, but as a result we simply classified him as a bit of a fraud.

Such trivial incidents were usually the basis of our assessment of teachers. Mostly they were male, and in their early twenties. We called them "Sir!", and as far as we were concerned they belonged in a different world. They were in authority, and there was no misapprehension about that. To confide in one was to court trouble; to delay brought a smack on the head; real disobedience brought a caning, done with due formality in the "study".

However, authoritarian as the prospects were, most of the teachers got along well with us, and we with them. Mr. Gausson, who taught English, Mathematics and Latin, had a flair for making outrageous puns (Latin lends itself to such manipulation). Not only did the ensuing student's groans help the lesson to go smoothly, but the puns were often an immense help in learning vocabulary. Mr. Hayward's forte was cricket. I do not remember learning anything else from Mr. Hayward, but he certainly taught me how to keep a "straight bat". (Regretfully this was to no avail, as I played no cricket after the age of fourteen).

There were a few women teachers, for the younger grades. Miss Pritchard, an angular but charming elderly lady, dispensed poetry ("The boy stood on the burning deck. . . ." and "It was the schooner Hesperus. . . . :"), and in recess would descend upon an unsuspecting boy (but only once), rattling a collection box for a contribution to the "lepers of West Africa". As we only received two pennies weekly, it was quite a sacrifice to give up one of them, no matter how worthy a cause. There was also a matron, to put iodine on cuts and bruises, of which we always had plenty, and to dose us with cascara.

The maintenance work, digging and unloading, was done by an ex-army sergeant-major, known as "Bonny". He was a cheerful, elderly, upright man, with a small waxed moustache. Well-liked, he was expert at removing nails from inside shoes.

We learned English, French (very good French, from Mrs. Arnold), Mathematics and Latin. Why Latin became a good subject for me, I will never know. I believe that the justification for taking classics lies in the theory that the only important function of school learning is to keep the young mind active; and the choice of subject is not important, as it will soon be forgotten afterwards. The theory has worked admirably in my case; I remember no Latin whatsoever.

In English we did grammar, essays, poetry and a lot of Shakespeare. The latter was taken very seriously, and in my last year we were involved with the production of King Lear, Julius Caesar and MacBeth, complete with casts of spearmen, soothsayers, citizens and retainers; a good time being had by all, as well as learning a lot of Shakespeare.

The school proper was in one central building. Adjoining were two buildings for the boarders. The dormitories consisted of large upstairs rooms, with upwards of six beds in each. By each bed was a chest-of-drawers, on top of which was a large china wash-basin, and a large china jug full of water at just above freezing temperature. Each bed had one pillow, and underneath was one chamber-pot of enamel. The window was always kept open at night. The matron was in charge when we went to bed, and each room had a senior boy, who was supposed to keep order. He usually encouraged one of us to "tell a story" once we were abed, as a way of forestalling any horsing around. Once in bed in the winter, it was just too cold to get out again, except for necessity.

In the morning, after as small a wash as we could get away with, we would assemble in the main building for "roll-call" by Bonny. Then breakfast in the main dining hall; either porridge, or some cooked-up staple such as fish-cakes; there was usually enough food, but I do not ever remember actually looking forward to it. After breakfast - lessons; lunch in the dining-hall; in the afternoon, more lessons, except on Wednesdays and Saturdays, when there was compulsory team sports. Supper was at 5 pm, after which - from 6 till 7.30 - there was compulsory supervised homework (called "prep"). After that, till 9.30, our time was our own, and it was probably spent in the basement locker-rooms, where it was possible to raise a little hell unrestrained by adults.

Everyone, no matter how unathletic, played the compulsory games, which were organized ahead of time, with a teacher as referee. In summer it was cricket, in winter soccer, in the spring English rugby. There was a first and second school team for each sport, and they had fixtures to play against other schools. If there were a home engagement, the rest of the school might be mustered to watch, and cheer (woe betide anyone who booed!). We also in the summer had track and field sports; and in the winter, indoors, boxing (which was also supposed to promote character). Never a fighter myself, I was nevertheless inveigled into several fights with other boys of my size, by the simple expedient of a teacher saying: "Oh well, of course, if you're afraid . . . ". I am quite sure that such fights did not improve my character one iota; whereas to have refused might have done so, as it would have fostered some independence of mind. (*That*, however, was not considered a desirable quality).

Mr. Arnold was married during my first year at Brighton, and his wife was like a breath of fresh air in the rather stuffy atmosphere of this English school. She was completely French, very attractive, wore the most alluring perfume, and spoke English with a fascinating accent. But she had a sharp tongue. She took a dislike to me early in our association.

The first time we met was when she came around the dormitories just before "lights out". This visit became routine. She would ask each boy in turn the question:-

"'Ave you been good?"

I considered this a rather silly question - we all did - obviously if one *hadn't* been good, one wasn't going to say so. So naturally we always answered in the affirmative; and continued to do so when the ritual was repeated each night. One night, however, I thought it might be interesting to see what would happen if one hadn't been good, so when she came round to ask her question to me, I answered "no!". The result was quite startling - I got all I asked for - and no amount of explaining that I *really had* been good was the slightest use.

"Oh, you bad boy, I don' want to speak to you at all. You lie down an' put on your head the bedclothes, you ver' bad boy . . . "

The trouble was that the treatment, once initiated, continued for about a week, and I think the stigma of being a self-acknowledged bad boy continued in her mind to apply to me, right up to my last term at the school; when, in French class which she taught, her sarcasm became too much to take, and I used a bad word. For this I was duly caned by her husband; but my conscience was clear. After all, I owed it to her to be a bad boy, and in the end I was.

My mother came to see me on one week-end, and took me down town for tea and buns. My father never came. But on one amazing occasion my grandmother did. Looking back, I think that what my grandmother wore when I knew her, and continued to wear in her old age, were the clothes which were fashionable when she was in her forties. After which she never changed her style. She wore an invariable color - black - usually with some lace around the collar, black shoes and long skirts. I imagine the stockings were black, too, had one been able to see them. There would also be a dark-colored hat, with a large hat-pin. I really appreciated this visit of my grandmother, as it was obvious to me what a lot of trouble she had gone to, in order to travel by train and bus to see me. She watched me play soccer, then took me downtown for the tea and buns.

For the boys whose parents didn't often visit, there were two kindly maiden ladies, the Misses Newton, who lived near the school. Each Sunday they invited two or three boys to "tea" at their house. The boys were selected by one of the teachers, and I went several times. For the first half-hour we were always polite, and on our best behavior, but the ladies were more genuinely interested than we probably knew, and usually succeeded in drawing out of us all the school gossip (which may or may not have found its way back to the faculty). They always sent us home stuffed to the gills with cream buns and chocolate cake.

Sundays were also the days on which we put on school plays, went on outings, and sometimes were entertained by a "lecturer". I remember slide shows on sheep-farming in Australia, lumbering in Quebec, and many conjurers. We also had a "lecturer" who professed to be, and may well have been, a cowboy. He came with a complete outfit - spurs, lariat and six-shooter (but no horse). We were fascinated , especially at the rope handling, which he did most expertly, so that lecture went on much longer than scheduled. In retrospect, the contrast between this man and the formal English school was quite ludicrous; but as boys we lapped up every minute of his time.

On the whole, whilst at that school, I was pretty confident, and as contented an animal as boys of that age group can be. I duly became a prefect in my last year (although - prefect or not - I got caned more times at aged thirteen than at any other time). In my last term, I sat for a scholarship to the higher school (college). I was the only boy who sat the exam that year; winning an Exhibition worth forty-five pounds a year, and being duly lauded for it. I left the school with mixed feelings; I did feel successful, having made a lot of friends there; but the prospect of entering a new school, where newcomers received regular hazing from the senior boys, was not too alluring.

SCHOLASTIC SERVITUDE 1.

"A schoolboy, while he is at school, regards his masters as a mixed assortment of tyrants and freaks".

– Stephen Leacock (1869–1944)

The year was 1930. The college was on the south coast of England. It consisted of a collection of stone buildings in extensive grounds. The original School house dominated; grouped about it were the headmaster's house, chapel, armouries, classrooms, science building, gymnasium, swimming pool (indoor), quadrangle, and - at the back - two playing fields. There were other grounds removed from the college containing more playing fields - when six hundred or so boys all play football at the same time, at least twenty five laid out fields are required.

Some students lived in the town, and attended school by day; but most were boarders, and they lived in "houses". The house system was supposed to foster "esprit de corps". Each house had its housemaster, who looked after the boys in it, appointed house prefects (senior boys who had powers of discipline, but only within the house), and generally stood "in loco parentis" (to use a dusty old expression). Team sports were organized as inter-house matches; a house could have teams A to F, which would play in leagues till the end of the term, with challenge cups to be won in all grades. Good players would also represent the school itself in the first XV (for rugby), or XI (for cricket). To be a member of such a team was, in prestige, equivalent to sitting at the right hand of God, and such illustrious boys (whose names were known to all) were regarded with awe by all the junior boys. In fact, the hierarchy of the school was based mainly on team sports. Academic prowess was not so highly regarded, although to be a member of the sixth (top) form did have a certain standing.

When I joined the school there were five houses. School A, School B, Chichester, Wilson's and Stenning. The latter two houses were outside the school grounds. Stenning - to which many of my friends from the former school went, and to which I was to go - was half a mile from the school. Each morning after breakfast we would take the necessary books, and walk up to the college in twos and threes. I made a trio at first with two boys who had formerly been at the lower school. Our way led past the rear of the Kemp Town Brewery, where at that time of the morning, the "mash" was being ejected from two large pipes in the wall into waiting carts below. There were always several carts, each pulled by a big Clydesdale horse, and we would often stop to admire the shining harness, and perforce savor the mingled odors of yeasty mash and horse-manure. If it is possible for smells to be portentous, then the odor of that latter commodity was certainly appropriate as an inauguration to my three year sojourn in that school.

Our housemaster's name was Tomlinson. He was a conscientious man and well liked. He had an invalid wife and two daughters. Later - when the house had been closed, and the boys had moved to Wilson's house - I used to visit him, and he always gave a warm welcome to his former pupils.

The hierarchy of Stenning House was interesting. The head boy was also the head of the whole school. His claim to fame lay in being the fastest quarter-miler (in track and field sports), and an unsurpassed wing-three-quarter (on the rugby field). He wore a gown, as all school prefects did, and he was one of the most stuck-up characters that I ever had the misfortune to run up against. His companion in the same study-room - one Blake-Whyte - was also a school prefect. He was the fastest swimmer in the school team (competitive swimming and water-polo were an accredited sport, though probably ranking second to cricket). He also gave the impression that he was on a higher plane than the rest of us plebeian boys. These two characters more or less ran the house as regards discipline.

The senior boys had separate study-rooms, two or three to a room. The rest of us kept our books in the common-room, where we had a desk and a locker. Homework was done by the juniors in the common-room, supervised by a prefect; the seniors studied in their studies. New boys spent their first term as "fags". According to roster, one would be detailed as "fag" for the day to one of the prefect's studies. This meant filling coal buckets, cleaning shoes, fetching, carrying, lighting fires. I do not ever remember getting a civil request to do any of these chores. Names were seldom used; the usual form of address of senior boy to junior being: "Boy!", or "You boy!", or maybe, "That boy there!".

In my first term as a new boy - thanks to the coaching of Mr. Hayward at cricket - I scored a hundred runs in one match (a century). I was duly summoned by the head of the house, and congratulated in the most condescending manner. This may have been the reason that I promptly stopped playing cricket, and enrolled as a competitive swimmer. I find it hard to write objectively about this school, even after more than fifty years have passed.

The headmaster of the school was a very well-respected clergyman. He was known to all as the " Chief". We all received a lecture from him on arrival, but from then, apart from the time he presided in chapel, we only saw him occasionally, at a class known as Divinity, where he was a very forceful teacher.

School classes in Britain were graded quite differently than here in Canada. The highest class was the sixth form, and only the brightest students got there, and in their last year. There were two "sides" - Classical and Modern. Classical (to us) meant Latin and Greek; Modern meant chemistry and physics. Thus there would be a IIIrd form (M), and a IIIrd form (C). For some reason which I couldn't fathom, it was found that if I went on the Modern side, I would be further behind than if I went on the Classical side; in spite of the fact that the only classics I had learnt thus far was Latin (I knew no Greek, nor Ancient History - which was part of the Classical curriculum). I am quite prepared to believe - now - that the reason I ended up on the Classical side was simply because there was a shortage of scholars who wished to study the classics; since I had won a scholarship, and had taken Latin, that was enough. I was relegated to the III(C) form, and spent my first week learning the Greek alphabet.

We slept in dormitories in the house, and - as in the former school - attended roll-call in the morning before breakfast. Being late for roll-call brought certain penalties. Twice late in a week, and one was awarded a caning by the head of the house; once late meant a "drill" on Saturday morning. The drills were taken by Sgt. Maj. Becket, a lantern-jawed ex-army warrant officer (albeit with a sense of humor), who stood in the middle of the quadrangle, while we - in gym clothing, and in column of fours - doubled round and round the perimeter, alternately raising our arms above our heads, and lowering them on command. The drill would last an hour, and some boys would take part voluntarily as athletic training. The caning by the head boy consisted of four strokes on the backside with a "swagger" stick. It was the recognized form of punishment, and we simply accepted it. In fact, a caning by ones housemaster, for a more serious offence, was much to be preferred to a "lashing" (as ones friends would sadistically describe it) from the head boy. The head boy at Wilson's House - to which I was eventually posted - was a large heavyset boy, who used to practice his art on a pillow, with a chalk line drawn on it.

Every morning before lessons, we attended chapel. The chapel was as big as a normal church, and it had a choir, organ loft and choir-practice room. Behind one of the choir seats was inscribed the name of one Richard Cotton - an ancestor about three "greats" removed. (Could this be the reason why I was sent to this school?)

The daily service was quite short, but on Sunday there were two Church of England services. I was in the choir from the beginning, as a treble; then, as my voice broke, as an alto; and finally as a tenor. We sang the hymns and canticles in harmony, and often had interesting anthems. Choir practice was on a week-day evening. The music master and organist was a Mr. Allen. I was to have piano lessons from him during my first and second terms; my mother sprang for the cost of them from her "dress allowance". I was thrilled with the idea, never having had formal lessons before; and enjoyed both the instruction and the practice (on Mr. Tomlinson's piano). There were definite signs that Mr. Allen's fondness for little boys exceeded that which would be considered justifiable as between teacher and pupil. As fourteen year old boys, our knowledge of such matters was quite sophisticated. However, no one that I remember was ever compromised (if that is the right expression), and Mr. Allen was a very good musician. Moreover he must have thought that I had some talent, as, in my second year he had singled me out to play a solo in the annual concert. Had I known of this at the time, it might have boosted my sagging ego; on the other hand, by then, the idea of playing alone on a stage, in front of the whole school, would have demoralized me completely. However, I caught diphtheria early in that term, spent a month in hospital, and was discharged a week before the concert, where I watched thankfully my good friend Oettle play instead. He was older than I, and went on to become a professional musician. I often wondered what became of him.

The teachers at this school all shared one quality - an inflexibility of purpose insofar as keeping order was concerned. The theory that if one can keep the little blighters in order one can teach them anything, was clearly the guiding principle. In a class of twenty or so boys, most of whom would rather be outside kicking a football than sitting in a stuffy classroom in a state of repressed silence, it was most important that the teacher be the dominant figure. So, it is mainly on account of their disciplinary methods that I remember these teachers, rather than through any knowledge gained by their teaching.

When Mr. Farnell, a mathematics teacher, observed a non-attentive boy in class, he would make no remark, but, as he walked in the aisle between the desks and came within reach of the boy, he would grab him by the hair, yet carry on walking and expounding. The unfortunate boy would perforce have to leave his desk and follow, bent almost double, much to his discomfort and the merriment of his peers, until the teacher ordered him back to his desk. In this classroom, it paid to have a short haircut. Mr. Hett, who taught classics, needed no gimmicks. He held the class simply by the force of a compelling personality. He was a good teacher - even now I remember most of the Greek History that he taught. He literally bullied me into diligent study, for which I respected him.

However, most of the teachers used corporal punishment, or at least the threat of it, as a disciplinary measure. One illustration of this remains vividly in my mind. My friend Oettle and I were in the same mathematics class in my second term. The teacher was one Davison - a first war veteran, who had apparently suffered shell-shock at that time. He was quite a moody person, as we all knew, and his moods could vary irrationally from being utterly charming to completely mean. There was, at mid-morning, a ten minute break, to enable classes to move to a different room. On this occasion, the class which included Oettle and myself did not have to move, so the ten minutes was for us free time. It was being put to good use by one set of boys, who were attempting to hold the door shut against another set who wanted to get in. With still a few minutes to go before lesson time, the door began to heave tremendously, at which time all the boys on the inside gave up the struggle, leaving only Oettle to face the intruder - who turned out to be a very wild Mr. Davison

He summoned Oettle to his desk. From it he produced a long, crooked cane, and straightened out the kinks in it. Then, after inviting the boy to bend over the desk, he gave him seven or eight strokes with all the strength of his arm. There was no sound from Oettle - or from anyone else. A dropped pin would have sounded like a rifle shot. Two minutes later, that amazing man looked up from his desk, smiled at Oettle, and said: "Hope I didn't hurt you too much. I'm afraid I lost my temper".

It is strange that so many bad memories of this school remain. Time is supposed to blur the edges, but they are still sharply etched. However, the fact that most situations were new, and therefore interesting, made us more or less content with our lot. I had, during the first year two good friends, also ex-students from the lower school. Had anyone asked us if we *liked* school, we would have considered that person quite out of his mind to ask such a question. When the end of term came, we were simply overjoyed at the prospect of release.

My own release meant spending time with either a paternal uncle (my namesake Hugh), an officer in the Royal Navy, who ran his household as if from the quarter-deck; or with an uncle on my mother's side, at Benllech Bay, Anglesey, where every day was a happy memory.

However, the holidays, whether strictly organized or free and easy, always came to an end; and I found myself returning, bag and baggage, to the school.

SCHOLASTIC SERVITUDE 2.

"It doesn't make much difference what you study, as long as you don't like it".

– Finley Peter Dunne (1867–1936)

Stenning House lasted for about three school terms after I joined the school. It was at least a quarter of a mile removed from the main school buildings; however, although the boys in the house were quite content with such isolation, no new boys seemed to be allotted to us - I was the last. As senior boys left, the number of pupils dwindled. Mr. Tomlinson was a popular housemaster, and when the house came to be closed, many of the pupils left the college altogether.

Although in most houses "hazing" of the junior boys was standard practice, this was not so at Stenning House. For two terms the junior boys, of whom I was the most junior, considered themselves fortunate in this regard. However, boys will be boys; and although ordinary horseplay is a perfectly healthy activity, it can and did lead to unhealthy extremes. On one occasion, the boys of the common room - looking for some excitement -- seized upon one of the juniors (his name was Ellis, and he had in a previous term been a good friend of mine). Ellis's treatment was to be carried to the basement in the midst of a shrieking mob of boys, stripped of his clothes, and to have black boot polish rubbed all over his genitals. I do not remember that Ellis was a particularly unpopular boy, simply that he was available. But after the treatment, I remember that he was not a very happy boy. The other juniors, including myself, fully expected that one of us would be next on the list of subjects. However, mob psychology seemed to dictate that the next victim should be Ellis again, and so it was. I noted that Ellis was not present after mid-term that year; presumably he had parents who were concerned.

That term was also the term when things started to go wrong for me.

In the initial academic stages, I had been reasonably content. Being in the III form (C), meant learning Greek, and continuing with Latin, as well as other more normal subjects. We read Xenophon, and Julius Caesar; and in due course I was promoted to the IV form (C). Here however, problems reared their ugly heads. As a "scholarship boy", it seemed that I was expected to excel - particularly in some subjects which actively turned me off. By that time I had found that, although Herodotus (whom we were now translating) was a most interesting man (I actually *liked* Greek), Livy and Vergil were a complete and utter bore. This did not please my teacher, Mr. Hett, who obviously regarded me as the ultimate challenge to his teaching powers. I remember sitting in his class (there were no more than eight pupils, whereas "Modern Side" classes had about twenty), actually in tears, while he railed at me for being - as he put it - intellectually lazy, yet well able to excel at the work which he prescribed. Eventually the threat was made that my scholarship would be taken away if I didn't smarten up.

So, a new era was ushered in. Extra work in the evenings in Mr. Tomlinson's study. No more music lessons, or practice. And since by all this I acquired the reputation of being a "swot", I found that I had few buddies. This whole episode may have been due to my having spent a month in hospital, which put me behind in my studies. I would have liked to think so.

Stenning House eventually closed. A great many of the inmates left the school at the end of that term. The rest of us opted to transfer to Wilson's House.

I think that before I describe the new house, I should say a word about homosexuality in the school. Although most adults, and particularly those who are wont to write books about boy's schools, would like to pretend that such a thing does not exist, to us the pupils, it was a very much alive subject. So without making any moral judgements, it is not untoward to describe its ramifications in every day school life.

New boys were always looked over carefully by the older boys, and usually the scan included an assessment of the boy physically. If he was merely good-looking, no particular notice would be taken. But if he were at all effeminate, that fact would soon become common knowledge. A handsome young boy was often referred to as a "tart", and, being in his early 'teens would be "adopted" by a senior boy, who would undertake to protect him from any harsh treatment by his peers. There were never, to my knowledge, any cases of sexual assault; but male masturbation was commonplace, singly or in pairs. I do not see that it could have been otherwise. In fact, if any person were to put the argument that such goings-on do not exist in all-male boarding schools, I would have to reply with one brief, succinct phrase - a phrase that I first learned at such a school, along with all the other four-letter words. Nowadays such words are in fairly common use, but I must say that they were in far more common use amongst us, the schoolboys of that era - as well as other language which had they heard it, would have turned our mothers to stone.

Wilson's House was situated about two blocks from the main school buildings, so-called because the housemaster's name was Wilson. He was known as "Bill" Wilson - or just plain "Bill" - and he was a large man with a broken nose. Rumor had it that he had been a heavyweight boxer at one time. His wife (known as "Ma Bill") was almost as large. I would not have relished three 3-minute rounds with either of them (Queensbury rules of course). Both seemed very even-tempered, and deliberate in their management of the house. I was never in trouble with them; on the other hand, I did not like the house or its inmates, and the few friends that I now made in school were boarders in other houses.

The house had a small library that I found quite stimulating. It was there that I made the acquaintance of Stephen Leacock, who as a Canadian writer was then well into his career. I had never read such an entertaining author, and what mainly endeared him to me was his uninhibited assessment of many of the British institutions which we, as scholars, were encouraged to regard as sacred.

By this time, our education in English literature included the reading of various novelists. I learned that - apart from the classical writers such as Dickens and Walter Scott - novelists wrote either good literature or "trash". The good novelists included:- Sir A.Conan Doyle, Rudyard Kipling, Joseph Conrad and John Masefield. We were encouraged to bring to class any novels of the day which we thought worthy of note, and these would be criticized according to the teacher's rating eg: E. Phillips Oppenheim - OK but trashy: Baroness Orczy (the Scarlet Pimpernel) - OK but a bit trashy: "Sapper", who wrote the Bulldog Drummond series - quite a bit trashy: Edgar Wallace - definitely trashy. As a matter of fact even some of Sir Arthur's writing was not very well approved. We spent two school terms wading through "Sir Nigel", followed by "The White Company", but it seemed that the Sherlock Holmes stories had some trashy overtones, so we did not read them. I did bring one of Stephen Leacock's books to class - "Nonsense novels" - but an indulgent smile was all the criticism it warranted, such colonial humor being hard for the establishment to understand.

By my sixteenth year, I had progressed to a study in Wilson's House, where four of us kept our belongings, did our homework, and occasionally cooked ghastly fry-ups of sausages, chips and baked beans on a gas ring, often to the accompaniment of music from a hand-wound phonograph.

The "hazing" of junior boys at Wilson's House was a regular and well-organized affair. About mid-term, one of the senior boys (not a prefect, as obviously prefects could not condone such activities), would call for a meeting in the gymnasium. The more senior boys who had studies, including myself by then, would stand on the floor of the gym, the junior ones would be herded on to the stage. There would follow a solemn harangue by the senior boy, to the effect that "the juniors are getting far to uppity, and need to have their arses kicked, and now they are going to get just that treatment". The procedure that followed might vary a little, but it mainly consisted of each junior having to sing a song, while the seniors pelted him with running shoes, old gloves, shin-pads and assorted brick-bats. The fact that the juniors invariably became more uppity after such "hazing" did not alter the routine, which took place each term.

Each year at the school, Mr. Allen produced a different Gilbert and Sullivan opera. The chorus consisted mainly of the boys who were in the choir, the female leads being played by younger boys, some with outstanding treble voices. The productions were very good, many of the teachers having good voices and being good actors. Mr. Hett, in particular, was a very believable Captain in H.M.S.Pinafore. But female as the boy sopranos might appear when singing, at the end of the song they were all boy, as they strode off stage with long rolling gait; to walk off like a girl with short mincing steps was an accomplishment few boys would attempt.

The headmaster of the school retired about this time, his place being taken by his deputy, Mr. Belcher, who is enshrined in my memory because of one lone incident. Shortly after he took office, he had occasion to summon the whole school to a meeting in the gymnasium, where he informed us that a very heinous crime had taken place. It seemed that a boy had had the temerity to leave the school grounds, and enter a public-house two blocks away, drink two beers, and then enter the college in time for classes. Mr. Belcher admitted that he didn't know the boy's name. However, when he found out who it was, it would mean expulsion, no less. We - the listeners - were quite relieved by this latter statement. I had received the impression that Mr. Belcher would have dearly liked to have the boy pilloried.

I do not think the life of a pupil in this school was any different from that in hundreds of other such schools in Great Britain at the time. There would be the same discipline, much of it administered by older boys, the same emphasis on team sports, the cadet corps, the physical training, the chapel routine and the classes - the latter being the only concession to actual learning; all the other aspects of school life being for the purpose of so-called "character-building".

At sixteen and a half years old, I left the school, having attained my certificate with six credits; these were the only credits I achieved. But there was also a debit.

This debit was a massive inferiority complex, which lasted for at least five years afterwards. The finishing touch to this state of non-confidence was provided by my ex-naval uncle, who decided that I needed "straightening out". This was done in true naval fashion (ending with the bone-headed recommendation, which could only have been made by a career service officer, that I should "clean my shoes more often"!). It left me with the feeling that I should apologize for presuming to exist.

When a person continues to regard Authority as not only correct but admirable, the result of continued rebuffs can only lead to a belief in his own ineptitude. Three years of this methodical shit-kicking, being alternately castigated and ignored, yet accepting the treatment as just, effectively turned me from a well-adjusted thirteen year old, to a demoralized seventeen year old. I was able to pass my University entrance exam, but was unable to look any discerning adult in the eye.

There is no doubt that the problems which I and many others acquired at this period of our lives were due mainly to a surfeit of Authority (with a capital A). We were not educated, we were indoctrinated. Defensiveness came later, and with it a duplicity of purpose, which would enable us to outwardly pander to Authority, while inwardly consigning it to perdition.

There are, of course, times when "Authority" needs to be consigned quite outwardly and openly to perdition. Although it has a place in our society, it needs to be reminded from time to time of its limits. Unfortunately during the school years, even if we had known of such a theory, we would have been quite unable to put it to use. We believed that our elders were in fact our betters, and it was not until some years had passed that we found out how false was that old adage.

3. WAR-TIME.

1939 RECALL.

"War is much too serious a matter to be entrusted to the military".

— Georges Clemenceau (1841–1929)

In 1939, in common with hundreds of thousands of other young people of my age group, I became embroiled in the turmoil of World War II. "Embroiled" is a somewhat misleading word, conjuring up, as it does, an image of sustained and even frantic effort. But there was the war, and I was in it, and I was thus able to corroborate that well-known maxim that "war from the point of view of the serviceman can be defined as long periods of boredom, punctuated by shorter periods of sheer terror".

There were some who, when the business was over, wrote books, such as Earl Burney's *Turvey* (a hilarious, down-to-earth book, that put army life in its proper perspective. I read it three times).

But most of us put the past six and a half years behind us, and out of our minds. I myself stopped even mentioning it in conversation, as the newer generation, who had not served in it, were apt to classify anyone who had done so as a dinosaur. Moreover, there were far too many challenges to be taken up in the post-war world to give much thought to the war.

More recently I have read a few other books of that time. I ploughed through all six of Winston Churchill's memoirs, and after so much time has passed, and before recording any of my own efforts during those years, I thought it would be wise to read a few other war books.

I came to the conclusion that in such books, realism varies inversely with the distance separating the author's location from the fighting front. In fact, the narrative of one particular incident, as described by two different authors, could result in two completely different accounts; one, detailing the actual devastation and the killing; the other, describing the "wonderful and glorious exploits". In the latter account is usually appended (with little comment) the statistics of casualties suffered by "our side" (the more wonderful the exploit, the heavier the casualties). Yet each of these statistics was a life - the life of a young man of my generation, cut off in his prime.

Probably the best book that I have read about those war years was by Farley Mowat (*"And No Birds Sing"*) - realistic, factual and with a message implied. I also read *"The Green Beret"*, by Hilary St. George Saunders* - with interest, since I had been in No 4 Commando. However, I have to classify parts of it as belonging to the second category of books I have described.

The first six months of the Second World War have been rightly labelled "the phony war". My own experience during that time was certainly in accordance with the label, but it was not without interest.

I had been employed since leaving school by a wholesale draper's establishment in the City of London. I was lucky to have a job at all, as it was the depression era. I cannot say that I liked the work, which I found boring in the extreme. However, I did not have either the drive, or the experience to strike out on my own, moreover like many young people of today, I had only a vague idea of what I wanted to do in life. In 1938 I had joined the Territorial (reserve) Army, so in September 1939, when war was declared I was one of the first to be "called up".

Much has been written about the attitude of the British people in the early days of the war, so I will not attempt to describe the complete feeling of resolution which prevailed in the country. It was shared by all of us - "us" being the young people of my age, of middle-class background (an all-embracing category) who had been brought up in the tradition that the First World War, in which their fathers had fought, had been a war to end all further wars. We were under no illusions about war. We knew about the horror of it, and the slaughter which had taken place. That it should happen again was unthinkable.

Armistice Day every year was a time to remember, and to mourn. On that day, church services were held with special hymns in every city, town and village; all had their cenotaphs; and at eleven o'clock throughout the whole of Britain there was two minutes of silence. All vehicular traffic - cars, trucks, buses - came to a standstill. In offices, restaurants, warehouses business simply stopped at the sound of the eleven o'clock siren, and did not resume until two minutes later. At the tomb of the Unknown Soldier, the Last Post was played, and for that two minutes London became a vast silent city. Every year of our youth we were reminded of the holocaust that had taken place between 1914 and 1918. We grew up with that concept, and it could never be discounted or forgotten. We were a nation utterly opposed to war, and our thinking was reflected in the attitude of our politicians, who not only advocated disarmament internationally, but implemented it at home by emasculating our armed forces.

Once a year we were thus reminded. For the rest of the time we teenagers went about our business. On Sundays we went to church to pray and sing praises to the Lord, whom I vaguely imagined as some huge cricket Umpire in the sky. I do not remember whether he wore several hats, as cricket umpires were wont to do, but he was obviously there to see that there was fair play amongst us (the British) and other (lesser) nations of the world. Much of the history which we had learned at school tended to reinforce the idea that the British had a special mandate from the said Umpire to see that such lesser nations behaved themselves. (Although it is only fair to say that our text books were no more biased nationalistically than the textbooks of other countries). Apart from the task of benignly presiding over the nations of the world, the Umpire had other more practical functions. This we learned at a very early age. For example, he was there to see that you didn't steal, or cheat, or lie, and would punish you if you did. The punishment was, of course, always meted out (on his behalf) by an adult. But the damnable part was that, even if you didn't get found out, He knew about it; and the burden of that guilty knowledge was almost as bad as the punishment itself would have been. . . . . In retrospect I think we were actually quite normal teenagers, but certainly religion, and all aspects of the British Empire, had played large parts in our upbringing.

During my last year at school, we had a class that met once a week, and was simply called "General knowledge". In fact it was a current affairs program, and a very enlightened one. Most sixteen-year-olds do not know or care much about world affairs, but we were able to appreciate the international scene, observe the rise of the Nazi party in Germany, and appreciateits repercussions on our own society. In the mid -30i's there was the Clivedon set, the Mitfords and Sir Oswald Mosley who believed that a good dose of Fascism was what Britain needed. We read about them in the newspapers; and although my friends thought (with schoolboy logic) that anyone with a name like Oswald just could not be taken seriouly (!), there were some that did so. In fact, as time went on, these factions ceased to make news (Sir Oswald was interned after war was declared), and when Hitler marched into one country after another, after publicly announcing that he had no territorial ambitions in any of them, the reaction of my contemporaries perforce hardened as we realised that war was inevitable.

So I have to record here something of the outlook of the people in Britain generally in those pre-war years. Hitler's policies brought about a decided change in attitude; from that of a peaceful nation to one of dedicated antagonism to the Nazi regime. We started to re-arm a scant two years before the invasion of Poland; and in May 1939, when it was obvious that war would be inevitable, I was one of the droves of young men who joined the Territorial Army.

RIGHT-OF-THE-LINE.

"From the sublime to the ridiculous is but a step".

– *Napoleon Bonaparte (1769–1821).*

Canada has its Reserve Army; in Britain there is an institution called the Territorial Army. Both consist for the most part of men who earn their living in other walks of life, but also give up part of their spare time to the service. Drills would take place twice per week, and on Saturday afternoons at a local drill-hall, and attendance at a minimum number of drills would be mandatory. There would also be a camp of two weeks duration in the summer, at which recruits lived under canvas, and took part in field maneuver. One was expected to use ones summer vacation time to attend this camp. Territorials were known as "Saturday afternoon soldiers". They would normally be drawn from men who worked in the vicinity of the drill-hall, which would be the headquarters of that particular Territorial army unit.

In the City of London there was an old established volunteer group known as the Honorable Artillery Company - commonly known as the "HAC". Its record of past service was impeccable, and it was highly regarded. It consisted of a battalion of infantry, and two batteries of Royal Horse Artillery. During the first World War, the infantry battalion, after suffering staggering losses, had been turned into an officers training unit. Since the headquarters of this regiment was in the City, the trainees consisted for the most part of well-educated men - stockbrokers, bankers and others whose normal uniform would consist of a bowler hat, briefcase and rolled umbrella. Accordingly, at the start of the second World War, the infantry battalion again became the nucleus of an officer's training unit, although the gunnery side of the regiment (which I myself joined) simply continued training as part of the normal artillery establishment of the home forces.

It is interesting to look into the early history of the HAC. There is no doubt that they were the oldest continuous regiment in the British army, by at least a century. They were incorporated by a charter of Henry VIII in 1537 - as the Fraternity of the Guild of St. George - at which time "artillery" referred to "long-bows, cross-bows and hand-guns", larger field pieces such as cannon having yet to make an appearance. An offshoot of the regiment was the "Ancient and Honorable Artillery Company" of Boston, Mass., formed in 1638, who fought on the side of the Americans in the War of Independence (and elected their officers). The Armory and drill-hall, where we paraded, was situate on the same piece of land in the City of London that had been granted to the regiment by Henry VIII. When we attended parades, we were treading on very historic ground. For four centuries this land had been used for drills, marches, guards of honor, ceremonial presentations, firing of salutes and other solemn exercises. However, according to Raike's history of the regiment*, up till 1878 (when the history was published) the regiment had only fought in one actual battle as a unit - the battle of Newbury (where I am glad to report they fought on the Parliamentary side). In view of what follows, I and my fellow recruits of the summer of 1939 had occasion to sympathize with those long-departed military men of that era. For 341 years they had endured the aforesaid "drills, marches, parades etc. We found that six months of such a diet was hard to take.

We were a mixed bag of recruits that summer. A shipping clerk, an accountant, a solicitor's clerk, a draper's assistant (myself) - a less soldierly bunch of men than could be imagined. The one idea that we had in common was to do something for the war effort, even to the extent of becoming soldiers.

Our drills consisted mainly of the elements of military behavior. We paraded in civilian clothes, marched and countermarched. Occasionally we would be given a lecture on the traditions of the regiment. Later in the summer, we reported to the drill-hall, and were issued with pre-war service dress, with ball-buttons, leather belts, light boots and peaked caps. Eventually we entrained for camp on Salisbury plain, having previously been assigned to squads as recruits. At the camp we slept six in a bell-tent with wooden floors.

It was here that we learned about fatigues. It seemed that if there were twenty men in a squad, then ten of them became employed on various domestic duties - cookhouse orderlies, latrine orderlies, HQ runners, ration parties, sentries and so on, in order to keep the other ten in training. So we spent just one day in two learning how to be soldiers. If anyone was sick then the ratio was less.

One unique aspect of this camp was that each tent was provided with a four gallon urine bucket, which was placed just inside the tent door. I don't remember any other camp where such a convenience existed, and I think the reason for it was to discourage men from making night-time treks individually to the latrine. One trek per tent was all that was expected, and the person who had to make the trek was supposed to be the last person to use the bucket before it was full.

Now it is amazing how soon that bucket filled up, especially when the six occupants of the tent had spent a couple of hours in the canteen drinking beer. Also, it is amazing how full a bucket can be, before it is actually judged to be full. It was not a pleasant job man-handling the brimming bucket in the dark for two hundred yards to the latrine pit, even without the added hazard of having a "Halt! Who goes there?" bellowed into ones unsuspecting ear-drum from behind by a sadistic sentry. When this did happen one night (not to me, thankfully), the carrier of the bucket was so startled that he spilled half the contents over himself, and was then so incensed that he dumped what was left over the sentry. The ensuing fracas resulted in the guard being turned out, and my unfortunate tent-mate under close - but not, hopefully, *too* close - arrest.

The day started at 6 am with Reveille, played on a trumpet. Lesser branches of the service might have bugles, but we, being right-of-the-line (the Royal Horse Artillery's prerogative) warranted trumpets. Very shortly afterwards - before in fact the sound of the trumpet had died away, would be heard the decidedly unmusical sounds of the Battery Sergeant-Major. I know of nothing that could banish the remnants of sleep so fast as that particular noise. In the next five minutes, all of us would be dressed and on the way to the cookhouse, where a brew called "gunfire tea" - hot, strong and sweet, was available.

Breakfast was cooked on a piece of equipment known as a "field kitchen". The design of these devices had not changed since World War I, in fact these particular ones might have been even older. Each consisted of a huge cauldron, like a giant washtub, capable of being towed behind a vehicle, and of having a fire lit underneath. It is amazing what a person will eat when he is hungry, and the physical exercise, to which many of us were unaccustomed, made us very much so. Porridge was usually the main course for breakfast. But, if boiled eggs were on the menu, they had to be seen to be believed. They came out black, blue and violet, and were hard enough to fire from guns. This must be Britain's secret weapon, we told ourselves, even as we consumed them for lack of anything better.

After breakfast there was a short period devoted to polishing brass - that is, belt buckles, buttons, cap-badges; and spit-and-polishing belts and boots. There were sundry other items all of which had to be kept bright and shining - frogs, scabbards, webbing equipment. Each soldier kept a small bag containing metal-polish, a button-stick, black and brown boot polish, brushes and rags for polishing. We also brailed up the tent, and laid out all of our kit for inspection. This had to be done exactly, in uniform fashion, and woe betide anyone who set an article out of alignment. The layout pattern was not always the same - it could be altered at the whim of the inspecting officer. We spent a *lot* of time in this way.

Morning parade was at 8 am. There were three trumpet calls. A "quarter warning", a "five-minute" warning - markers would be set out at the five minute warning - and the "ON PARADE" call, when all soldiers who were not on fatigues converged on the markers at the double.

There was always an inspection at this parade, and occasionally it would be done by an officer, also a Territorial. After the ritual of handing over the parade to this officer, the sergeant-major would take up a position three paces behind him, and the officer would *saunter* across the parade-ground to appraise at leisure the rigid ranks; with the sergeant-major instantly ready to make note of any critical remarks. Our own critical remarks had to wait until the parade was over; but I remember thinking how gratifying it would be if the officer's pants fell down, or if he slipped on a dog turd - after all who was he to saunter . Had we ever dared to do so, we would have been put on a charge for behaving in an unsoldierlike manner, and spent a week cleaning the latrines.

Although we had some instruction at that camp in such subjects as gunnery, map-reading, small arms etc. from regular army instructors, the disciplinary training was left to our own non-commissioned officers. The psychology of this training is interesting, and as a subject would fill a book. The phenomenon whereby a group of civilians, from different walks of life, and who have nothing particular in common with each other, can be transformed into a disciplined body of men, has to be categorized as brainwashing. Maybe "indoctrination" would be a kinder word. Such indoctrination could even have been good for us, in that it made us more outgoing, and developed leadership qualities. Those who bucked the system, however, fared badly. They ended up in the "glass-house". A place to be avoided.

After the first few days of recruit training, we found that we did have something in common with each other, after all. That was the unanimous opinion that Sergeant-Major B . . . . . . was a complete and unprintable bastard. This probably formed the basis of our eventually becoming welded together into a coherent unit - the fact that we were combined in our hatred for this gimlet-eyed monster, who could spot an unpolished button at a distance of one hundred yards, and verbally demolish the owner of it from the same distance. We were all fellow sufferers. We soon found out , however, that if we did smarten up, then we did not get reprimanded, and occasionally there might even be a word of praise for a good turnout, or a well-executed movement on the parade square. The next phase took place after we had met the S.M. in the canteen, off duty, when he bought us a beer and chatted; and we came to the conclusion that he wasn't such a bad fellow after all. The last phase soon followed, when we began to congratulate ourselves that we had *him* in charge of our squad, and tell ourselves how much better, and smarter, we were as a unit, compared with other squads commanded by lesser NCOs. This, of course, was the beginning of the end - we no longer thought of ourselves as civilians any more.

At the time I joined the HAC, the word "horse" in Royal Horse Artillery had become merely a concession to tradition (although the ultra-smart K Battery of the regular army did keep their horses and 13 pounder guns for ceremonial purposes). We had no horses, although I remember that whips were still issued as part of our uniform, to be carried when off duty (possibly in case we should come across a stray, off-duty horse). Our guns, however, were still the 13 pounders that had been used to effect in the Boer War, and had been more or less obsolete ever since. When the Second War was declared, these guns were withdrawn from us.

So, on the third of September 1939 we were mobilized, and both horseless and gunless, transported to the outskirts of London, where we were housed in empty blocks of apartments as billets.

Elstree with its Pinewood Studios, was just around the corner from our billets, which were known as the Hollywood Apartments. But no glamorous movie stars brightened our days, instead we had regular visits from various high-ranking officers, who stalked along our frozen ranks exuding a different kind of glamour.

In time, when the higher brass got tired of inspecting us, we were put to work on defensive measures. We filled sand-bags, put up barbed wire and - in view of the current concern about a possible fifth column in Britain - we were used for guarding VPs. "VP" meant vulnerable point. There were many in Britain, and all could have been prime targets for sabotage, had there been a serious fifth column.

One of these interludes, which I remember with interest, was spent on sentry duty in the Holborn Tunnel, in the City of London. Between the railways going north out of London, and those going south, there was but one connecting line, which went through this tunnel. It was a busy section of line. Since the smoke and coal dust would have spoiled our uniforms, we were issued with dark civilian overcoats (with wide shoulders and wasp waists). With slung rifle, tin-hat and gas mask at the alert position, we really did look like "Fred Karno's Army". Our sentry posts extended along this tunnel, one post for every hundred feet. In between were braziers - large buckets with holes in them, in which was burned coal. It was cold. We were not supposed to linger near the braziers. If we ran short of coal, we would shout to the fireman of a passing locomotive (they were passing all the time):- "Give us a lump of coal, mate!" - and a lump of about five cubic feet would land with a crunch at our feet.

We never caught any saboteurs. The orderly officer, on the other hand, gave us a lot of trouble. Since sound travelled well in the tunnel, you usually had an idea when he was doing his nightly rounds. You waited until you heard the adjoining sentry challenge ("Halt, who goes there?"), which told you that he would arrive next on your own beat. You then waited until he was about twenty feet away, then challenged him with every ounce of lung power. Timing could be important. If you made him jump, it was one up for you. But if he surprised you, and you didn't hear him coming until he was right close to you, you would be in for trouble. Of course, the procedure did not end with this challenge. The answer would come:-

"Friend!"

Then - "Advance, friend, and be recognized!"

"Halt!" (-five feet from your bayonet)

After which the officer would shine a flashlight on his face, to let you off the hook. Had any self-respecting saboteur taken the place of the orderly officer, we would have been as dead as mutton in short order; as we were to realize at a later date.

We had our share of characters in the Battery. One, whose name I have tried hard to remember, coined a very descriptive expression: "It's a fuckaroo, mates", he would say. It was an expression that could be used to describe most situations, and soon was in universal use.

Another VP which we guarded about this time was Croydon Airport, with its squadrons of Gloucester Gladiators. This was before the Battle of Britain, We had sentries all around the perimeter wire of the airport, so there was a large guard on duty every night. On this particular occasion, the guard commander who had the misfortune to have me as one of his squad, was a man named Martin. He was a personable young man, fairly popular amongst all ranks, and - this is what made him different from most NCos - he was the son of a minister. Because Sergeant Martin was different. In company of men whose every second word was profane, he seldom swore at all. In fact we respected him for it, as he managed to chivvy us around, and get things done without a single bad word.

It was usual for the old guard to parade at 7 am, before being relieved. However, at this parade one had to be properly dressed, and with rifle "clean, bright and slightly oiled". This particular parade had caught me unawares, and I only just made it to the rear rank as the sergeant called us to attention. I could see that I was a marked man. To make matters far worse, when it came to arms inspection, I found that my rifle was minus a bolt (I had been cleaning it when the order to "fall in" was given).

The two front ranks were standing easy, and several had turned to watch, when Sgt Martin arrived in front of me. The outcome was predictable.

"Cotton . . . . ", he muttered, "Cotton . . . . "

I could see that he was trying to find a suitable word, and eventually he found it.

"Cotton," he finally exploded, "YOU'RE A--------ING --------- !!"

Both noun and adjective had sexual overtones, and in civilian life these would be fighting words. But in the army, they were considered normal behaviour when used by a non-com to a lowly gunner (such as me); and. from the roar of laughter (there was even a weak cheer) that greeted this announcement, it was obvious that this was no ordinary rebuke. A milestone had been reached for Sgt Martin. From that day on he was a changed man, and started to swear like everyone else. As for me, the catalyst, I enjoyed temporary notoriety, and had a hard time handling all the free beer that came my way that day.

Sometime in the early part of 1940 we again moved; this time to the south coast of England. Here a proper training program was instituted, and there were rumors that we might eventually become operational. When we acquired guns, shortly afterwards, the rumors intensified. The guns were new twenty-five pounders, then the standard field artillery weapon, and there was a new type of tractor, lightly armored, with four-wheel drive. We began to train in earnest with the new equipment, moving in column about the English countryside, bringing guns into, and out of action. Map-reading, gun drill and driver-training became the order of the day. Unfortunately we were not very efficient, as we soon found out.

One day a small team of regular army instructors from Tidworth descended upon us. There was an "I.G." (Instructor of Gunnery)) to instruct the officers, and an "Ack I.G." (Assistant Instructor of Gunnery) to instruct the other ranks. Instruct us they did. We must have learned more in that brief encounter than in the whole previous six months. There was a subtle difference in the training - the emphasis was on efficiency; we were to become gunners, not well-drilled zombies. There was a complete absence of that commodity known as bull-shit, which up till now had played such an important part in our army lives.

After a short period the instructors returned to Tidworth, to make their reports, the content of which soon became apparent. The new equipment was withdrawn, and given to another unit which went to France in our place. It was left behind there when the German offensive of 1940 was followed by Dunkirk. In fact, it was all one big fuckaroo!

For us, it was back to square one. But when recruiting began for the Commandos (then a top secret organization), there was no lack of signees from our outfit. The old maxim about never volunteering for anything could hardly apply to us, as we had nothing to lose.

COMMANDO.

"War never slays a bad man in its course, but the good always".

– *Sophocles (495–405 BC).*

In view of the traits which have become associated with the word "Commando" of late years, some explanation is needed to describe how and why these troops came to be formed in World War II; and to dispel any ideas that we were either supermen or early prototypes of Rambo.

The idea of Special Service troops in the British forces was not new, although radical. The forerunners of the Commandos were the Independent Companies, formed at the start of the ill-fated Norwegian campaign in the spring of 1940. They were to be troops whose movable base was a ship, so that they could take part independently in amphibious operations. After the debacle of Dunkirk, and its consequent effect upon morale in the army, the "powers-that-be" (the CIGS and Winston Churchill) decided that a new arm of the service should be created, one whose sole purpose was agressive tactics, and could operate independently of the cumbersome regular army. It was to come under the direction of the Chief of Combined Operations (a new post), and to consist of picked volunteers, who were already trained soldiers, and who would be specially trained in amphibious operations.

The name of "Special Service" troops, owing to its similarity to the "SS" units of the German army, was dropped in favor of the name

"Commando". The original Commandos were irregular units of the Boer forces, in the Boer War of 1898. They consisted of mounted settlers who normally tended their farms, but who, if called upon to fight, could rendezvous as a well-equipped group at short notice, carry out a raid deep in enemy-held territory, and afterwards disperse before a counterattack could be organized.

Making raids in enemy held territory was the first function of our Commandos; in later years, however, they came to be used as the spearhead of all amphibious landings.

Recruiting for several of the units, including No.4, started in July 1940, and by the end of the year there was a Special Service Brigade in existence. Royal Marine Commandos were formed also, as the concept grew. Shortly afterwards, Admiral-of-the Fleet Sir Roger Keyes (the hero of Zeebruge in the First War) became Chief of Combined Operations. He was small, old and bird-like, yet at sixty-eight years he radiated vim and vigor. "A proper fire-eater" was our assessment of him. He was a good leader for a new concept; and it was plain, even to us lowly soldiers, that he would need all his leadership qualities. He had to contend with the established brass on the General Staff when planning operations, who would look with grave suspicion on most new ideas. Fortunately, Winston C. was behind us - he had spawned the idea - and he wasn't going to see it bogged down; so we continued to have a "strong" man as the Director. When Sir Roger had to step down, Lord Mountbatten took over, and finally General Laycock, who had originally raised and commanded No. 8 Commando at the beginning. So much for the undercurrents of politics, of which, strangely enough, we were aware; we thought it quite probable that the regular brass did not approve of us.

We of the HAC, stationed at Swanage on the south coast of England, were interviewed by the Colonel who was to become our commanding officer. The main qualifications for acceptance were that one was physically fit, a trained soldier and keen enough to volunteer. Ability to swim, and familiarity with small boats was an asset.

We joined No. 4 Commando which was concentrating at Portsmouth, and for maybe two months, while we "shook down" together, this town became our base. All of us from the HAC were formed into the same troop. We had a captain as troop commander, and two subalterns; We had also a troop sergeant-major, and NCOs from other units. We soon found that a great many of our co-volunteers were regular soldiers. Most had been at Dunkirk, and many had served in India and overseas before the war. No barracks were to be allocated us from now on. We received a "subsistence allowance" of six shillings and eightpence per day, which was to be used to find our own accommodation, and meals when at base. How and where we spent it was entirely up to ourselves; what mattered only was that we should appear on parade whenever and wherever required, and properly equipped.

At the beginning, of course, there was the odd man who decided that this lack of regulation was too good to be true, and took advantage of it. Such were quickly "RTU'd". This meant "returned to unit", and it could be implemented at the request of the Commando soldier at will, or - if he was considered unsuitable - by his commanding officer. But for the rest of us, it was as if we had just become of age; we were actually being expected to think for ourselves, an activity that had seemed to be lacking in our last billet.

This was the first step in fostering independence. If we were ordered to parade in a place five miles away, in battle order, we did so, and on time, and with our equipment clean and in good order. There was no way that we would abuse this privilege. Other advantages quickly became apparent. Since we lived in civilian billets, there were no fatigues to be done, and all our time could be devoted to training. We learned intensively about small arms; we took them apart, and reassembled them in the dark. We fired them on the ranges. We trained with the Navy in small boats. We did night marches, night operations, night embarkations; also physical training and unarmed combat; lectures on map-reading, living off the land, first aid, demolitions, "careless talk" - the two months passed quickly.

I had teamed up with two other ex-HAC men. One of them - Bill - had been the Battery trumpeter. He was tall, well-built, very handsome and sported an elegant moustache. Most girls took one look at him, and, figuratively speaking, fell over backwards. The other lad was a different type, but we seemed to get along well, so we sought lodging together. Bill had been fairly well-off before the war. He owned a Railton Straight Eight, one of the ultimate sports cars of the day. He used to drive it at ninety mph on the Portsmouth road - but since he could not get petrol (due to rationing), he had been unable to use the car much. However, with the advent of civilian billets of our own choosing, he soon wangled a forty-eight hour pass, went up to London and drove the car back to Portsmouth. From then on, our outings depended on how well he could bribe some service driver to look the other way, while a few gallons were siphoned out of his tank.

Bill was not satisfied with having a landlady. In short order, he had the three of us installed in a three-bedroom house in the suburbs, which we rented through a real-estate agent. We cooked our own meals for a while; but it wasn't long before Bill had made friends - and I mean friends - with the two young housewives next door on either side (their husbands were in the Navy, and from home). If we were home in the evening, one or both would come over to cook, and then stay for a party. We certainly lived it up in Bill's company, and the six and eightpence per day each did not begin to cover expenses. Brandy and soda was his drink if we had any money, and his capacity was great. We were broke a great deal of the time. Bill quite often used to go out at bedtime, and return in the early hours of the morning; he was obviously well-thought of by both our neighbors, but I was never quite sure which one of them he was sleeping with, although it might have been both. He was what we would call "a bit of a lad".

When the invasion scare was at its height, the Commando was assigned an unusual role. The thinking at higher levels was logical enough, that is that if the enemy were to land on Britain's shores (which at the time seemed quite a possibility), then all roads in the neighborhood of the beaches would be completely congested, so that any mobile reserve (such as ourselves) would be unable to make fast progress to the battle-front. We were therefore to be equipped with bicycles, so that we could move quickly through the stalled traffic. The bicycles became available at once - heavy machines with back-pedal brakes - and so that we could get used to operating as a "flying squad" (some of us had never ridden a bike before), each troop went off for a week's tour through the villages of south-west England.

I must say that it was the nearest thing to a holiday I ever had in six and a half years of war. A fifteen-hundredweight truck accompanied us, containing blankets, a cook, cook-stove, rations and one of the officers. We bivouaced at a predetermined place, picked from the map each morning when we started out. The officers rode with us. We were dressed in battle-order, but with full packs and slung rifles. A flat tire meant that the rider dropped out from the group. After fixing the flat, one had to ride hard to catch up. Fortunately, there were pubs all along the way (in case one got thirsty), and the odd passing truck going slow enough to be used as a towing vehicle (and drag one up the hill instead of walking). We returned after a week, all expert riders. However, the invasion scare was very soon over, and then we were under orders to move.

We entrained for an unknown destination, which turned out to be the town of Ayr, in Scotland. Bill had to go to London again, to put his car in storage. At Ayr we were photographed as a whole unit. Also we got a new C.O. Not that any of us objected to our present CO who had recruited us. He had been a very enterprising and straightforward leader, albeit with a low boiling point. (He castigated Bill as a "bloody idle sod" on one occasion, but we forgave him for that. Bill *was* a bloody idle sod).

Shortly after we acquired the new C.O. - a Lieut.Col.Lister, who, it was said, had once been the heavyweight boxing champion of the army - we moved to Troon, a few miles away from Ayr, and once more dispersed to civilian billets. This was Robert Burns' country. Local villages all bore names familiar to readers of his works (I was an avid one) - where dwelt the subjects who inspired his immortal verses (in 1942 we were to find the lassies bonny, too !).

I teamed up with two other fellows this time - I couldn't afford living with Bill, nice guy though he was. Troon was very convenient to the port of Glasgow, and beyond, where the highland sea-lochs, and steep mountains would be used as our training grounds. After a while our troop was split up, and several men (Bill included) returned to their units. Once a man was RTU'd, one would lose touch with him, and I often wondered if the rumor were true, that Bill had met his end at the hands of a jealous husband.

Troon welcomed us. It became No. 4 Commando's base for the whole war. I remember it with nostalgia. There were few of us who didn't have a girl-friend in town, and several of us married local girls. I lodged with George Ivison, another ex-HAC man. George, who came from London, became a particular buddy; six years older than myself, which was unusual as most of us were in the early twenties. We seldom disagreed, except maybe over a decision as to which pub to visit of an evening. I don't believe there were any pubs in the neighborhood of Troon, and nearby Ayr, that we weren't well acquainted with.

Training in combined operations continued. We received special issues of equipment - toggle ropes and fighting knives. (The green berets had not yet made their appearance in those early days; they were to come later). The prospect that we might find ourselves lying, gut-shot, on some wind-swept beach, as happened only too soon at St. Nazaire, and Dieppe, was not one that I remember any of us contemplating, although we were well aware of realities. That there would be casualties we didn't doubt, and in this we were no different from the "PBI" (poor bloody infantry).

Near the end of April 1941, we embarked at Gourock, together with No. 3. Commando, aboard two landing ships, the Queen Emma and Princess Beatrice. These were two former Dutch cargo ships. They would do about seventeen knots, and had been converted for use as Landing Ships. On the top decks were carried auxiliary landing craft (LCA's), and there were large mess-decks below to hold the troops. We slept in hammocks slung above the mess-decks. As was normal, we knew not whither we were bound. The chances could be that we would spend the next two weeks practicing landings in Loch Fyne. However, things turned out differently on this occasion. Our two ships rolled and heaved their way around the Hebrides, to find stable anchorage in Scapa Flow. A few days were spent here, during which time we were given a conducted tour of H.M.S.Nelson, courtesy of the Navy.

We were then briefed, issued with special warm clothing, and set sail for the Lofoten Islands in northern Norway. We had destroyer escorts on each flank, and - it was rumored - the Nelson, Rodney and other battle-waggons following. The Tirpitz was known to be holed up in Narvik harbor; we were to be the cheese in the mousetrap, to lure her out so that our Navy could deal with her. Apart from this role, our other purpose was to make a surprise landing at Svolvaer and Brettesnes, destroy various installations and take prisoners in the process.

The raid turned out to be a complete success, although (luckily for us) the Tirpitz would not be lured out. When we withdrew, we took with us two hundred odd prisoners, various "quislings", and a large number of volunteers for the free Norwegian forces in Britain.

I myself was monumentally sick for the first two days at sea; however, a "geordie" corporal brought me a plate of the greasiest-looking stew I had ever seen, and stood over me while I ate it. Whereupon I at once got up - cured. I can recommend the treatment. We launched the landing craft in the dark of the early morning, just off the Lofoten Islands. It was colder than I'd ever known, and as we approached land, the dawn broke to a brilliant blue sky, with snow-covered peaks in the background. The surprise was complete. Every item of intelligence which had been supplied to us was accurate. The visiting professor from Germany, who was scheduled to arrive that morning to give a lecture, showed up on time as scheduled, and was put in the bag.

Our LCA landed at a wharf, where we found a German freighter docked. We rousted out the captain and crew (the former in his nightshirt). Our troop commander then nailed up a photo of Hitler - which he had found in the cabin - and proceeded to shoot it full of holes with his pistol. The captain, however, was shivering so much either from cold or fright, or both, that the symbolism was probably lost on him. We took him and his crew with us back to the main wharf where two Germans, the only casualties of the raid were laying. Sergeant White had had his Bren gun on the wharf, and although we had orders not to fire unless specifically ordered, "Ginger" (still mindful of Dunkirk) was too eager, and shot up the rowing boat in which the Germans were trying to escape. I seem to remember that, on return to our base, Sergeant White became Private White; independent though the Commandos might be, discipline was strictly enforced.

We made an uneventful voyage back to Gourock, where secrecy had been lifted. The press were waiting. In the newspapers we were pictured waving from the ship, with captured swastikas draped over the rail. This was welcome news to the general public, who had been exposed since Dunkirk to a steady diet of gloom and austerity in their news stories.

After Lofoten, training became more intense, and it soon became obvious that some large-scale operation was in the wind. I had always regarded a route march as a pretty boring form of exercise. Most soldiers did, so singing was encouraged, and talking allowed - usually we marched "at ease". Any girls showing an interest in the passing column would be subjected to a barrage of whistles, or such outrageous remarks as: "'ad it in lately?". However, the forced marches which we now started doing were quite different. We covered eighteen miles in three hours, in battle order - with no rests at all, and no time to whistle at girls. Soon we embarked on the Princess Emma again, and spent weeks climbing the hills in the Scottish highlands. Lord Lovat was second-in-command at the time. George did not take kindly to his lordship, who climbed the hills clad in a mackintosh coat, and carrying a "swagger" stick, which he flourished from time to time; on the other hand, we - George and the rest of us - strode dutifully upwards carrying on our backs about sixty pounds of assorted equipment, and with the added spice of having live ammunition fired over and around us, for acclimatization purposes).

So, in the prime of our fitness, we again sailed away in the Queen Emma. This time, however, we only got as far as an anchorage close to the island of Arran. As near as I can remember, it was October 1941. Sir Roger Keyes came aboard, and in fiery speech of farewell, urged us to "keep our weapons bright", no matter how frustrated we might feel at not being able to use them. From this we deduced - we had become quite adept at reading between the lines - that (a) the operation for which we had been training had been cancelled, and (b) that Sir Roger was stepping down as Chief of Combined Ops. We were right on both counts. All ranks were given two choices: to remain in our present Commando, or to go forthwith to the Middle East as reinforcements for the three Commandos in that theatre of war (Layforce - and they had need of them). There was much deliberation, during which all of us ex-HAC men resolved to stick together. However, in the final analysis - as could be expected - half of us ended up going to the Middle East, and half stayed as we were. George, Reg, myself and two or three others stayed. Since the proposed operation had only been postponed, rather than cancelled, we were then landed on the island of Arran, for security reasons, and billeted there. The rest sailed.

Arran is a very small island. It is very rural. It was also - to our chagrin - dry. However, F troop were housed in comfortable billets. We stayed there until after Christmas, when the operation was considered officially defunct. Again we trained - marches, demolitions, map-reading - we even climbed Goatfell, hardly a mountain by Canadian standards, but with a spectacular viewpoint. But this time, the training for me had lost its savor. We were obviously just putting in time. For diversion, in the evenings, we played a horrendous game called three-card brag. It was an old army card game, and, as could be expected, the old sweats won all the money. This hardly mattered as there were no pubs in which to spend our pay.

We eventually returned to our old base in Troon. George and I returned to our old civilian billets. So many families in Troon "took in" a Commando, that this town had become like a second home to us. I took a transfer to the Signal section, thinking that it was time I used my head for a while. But the feeling of uselessness grew, and in early 1942 Reg and I both decided to apply for commissions. We thought that, maybe, if we passed the interview, and were considered "officer material", we might find ourselves useful in that capacity. We did, in fact, pass our interviews, and both went on to become officers; although I have to admit that the same feeling of uselessness, which pervaded my first six months in uniform, was present on occasion after I was commissioned; and one of the few times during the war when I felt really free of it, was the time spent in No.4 Commando.

I should here mention, that after Reg and I had left, No. 4 Commando took part in the Dieppe raid, at Bruneval on the left flank. Their part in the operation was judged to be an "unqualified success". Their losses were eighteen per cent. On the other flank, No. 3 Commando lost a hundred and twenty out of two hundred and fifty men. One of those wounded and taken prisoner was George. I did not see him again until it was all over in 1946, and he was home with his wife, and small daughter whom he had never seen until his release. When released from prisoner-of-war camp, he had weighed no more than ninety pounds.

As for the later exploits of No. 4 Commando, their history is well documented. It makes very sad - or glorious - reading: according to ones point-of-view.

H.B.C. at Cap Matifou 1942.

1939 - 1946.
(The only surviving photos)

Kitty in Baghdad 1942.

No 4 Commando, Ayr, Scotland 1941

"She says she IS Lili Marlene"

Monument to the Commandos in WW II, Loch Ailort, Scotland.

MORE WAR YEARS.

"A brass-hat is an officer of at least one rank higher than you, whom you don't like, and who doesn't like you".

– Kenneth Clairborne Royall (1894–).

"Temporary gents" was a current expression used to describe wartime soldiers, such as myself, who aspired to become officers; and there must be hundreds of us who remember the chain of events which led to the granting of the commission. There were often some quite droll moments.

After having had several interviews with senior officers, none of whom apparently expected me to do more than stand stiffly to attention and say:- "Yessir!", I finally went before a selection board, consisting of three high-ranking officers, one a major-general. These gentlemen bade me sit down (I sat at attention, of course) while I answered their questions (most of which the other interviewees and myself had anticipated). Apart from the:- "Where did you go to school?" one, which they didn't need to ask, as it was all written down for them on my file (but which I suspect carried with it much "old-boy network" potential) - the main topic of conversation was on the subject of: "Why do you want to become an officer?" This took up several minutes, in which my own rehearsed answer was duly trotted out. After which the general almost floored me by asking: "Why did you join the Commandos?". My reasons could have been partly patriotic, but if they were I was far too inhibited to come out and proclaim them openly. Momentarily, it occurred to me to make a joke of it and say: "Because the girls think we're wonderful, and we get lots of free beer . . . .", but fortunately I dismissed *that* idea as inappropriate, contenting myself, after a pause, with declaring: "Because I wanted to fight . . . ", and that must have been the right answer, because I was hired! Just one week later I was posted to Carlisle, to an artillery training depot, to take a pre-OCTU (Officer Cadet Training Unit) course, where we didn't do any fighting, but instead did a lot of "square-bashing" - precision drilling, marching, and countermarching on the parade square.

We were, of course, no strangers to "square-bashing". In this instance, however, there was a difference; we were the ones who took turns in handling the squad, and ceremoniously bawling out any of our buddies who needed smartening up, knowing that they would duly return the favor when it was their turn "out front".

So here I should introduce Sergeant Powell, an outstanding and dedicated drill instructor.

When Sergeant Powell stood at attention, he was as straight as a ramrod, his battle-dress was immaculately creased, his belt perfectly blancoed, his cap-badge dazzling, his boot-toes like mirrors; in addition to this, his chest stuck out, his bottom stuck in, his thumbs were to the front (and lined up with the seams of his trousers), the toes of his boots were inclined at an angle of forty-five degrees, his heels were together, his head was level, and his eyes looked straight to the front, and - what more can I say? Sergeant Powell was a perfect soldier, all five foot nine inches of him; a fact which not one of us would have doubted, had not the sergeant made this statement himself so frequently, and exhorted us daily to the effect that, by the time the course was finished, we were all going to be like him.

This was the standard treatment meted out to recruits, but after two years of war it had ceased to ring true; moreover, none of us were recruits and had been considered 'officer material'

The situatioin was not without humour. Marching (according to army regulations) requiured that the arms be swung straight forwards and backwards with each step. The straightness was important, as some soldiers - in an excess of smartness - tended to turn their arms inward on the backswing, partly behind their back. This was frowned upon by all good drill instructors, and referred to colloquially as 'wiping your arse'.

I mention this because, when on the parade ground, our squad would occasionally do this movement deliberately, just for the hell of it (to see what would happen) and, sure enough, sometimes from at least 400 yards distant, would come Sergeant Powell's raucous bellow - 'THAT SQUAD THERE - STOP WIPING YOUR ----------ING ARSES' !!

Our squad, recalcitrant though it may have been at times, had their hearts and minds pointed in the right direction. We were from all walks of life, non-coms (we even had a sergeant-major) and we developed quite a bond of camaraderie. Many of the personalities are still etched in my memory. Sad to relate several were to lose their lives later, on active duty.

It is just possible - but only just - that the course did us good. In fact we all felt somewhat sorry for the sergeant having to go through this procedure for a group of civilians-in-uniform who, when they were off-duty, assessed the training as just another brand of unnecessary bullshit. He did have an unenviable job.

We all passed the course.

At Llandrindod Wells, the OCTU for Light Anti-aircraft Artillery officers, the atmosphere leaned towards the academic, rather than the martial, although we obviously were still in the service and marched in squads from one classroom to the next. But we were there to learn, and so we did. All about the Bofors light ack-ack gun, ballistics, aircraft recognition, map-reading, military law, man-management, and all the dreary details of the organization of that king of bureaucracies - the armed forces. We wore a single white band on our shoulder strap, to indicate officer-cadet, while retaining our current rank.

This period of war was one of forthright effort. Rationing was just one of the innumerable hardships which the civilians bore, and was looked upon as a necessary evil. We were regimented beyond belief, but we accepted it. We were exhorted to dig for Victory, pray for Victory, save newspapers, save bottles, save fuel, save old iron, keep fit for Victory, buy Victory bonds. We were told that careless talk costs lives, to keep our bowels open and our mouths shut, and innumerable slogans were presented to us visually every day, in railway stations, on buses, hoardings, subways and highways. There were no motorways in those days, the closest approximation being the Great West and Great North roads, which, in combination, with the concept of driving on the left, must have caused our allies the Americans untold frustration. For the USA was now in the war, the Battle of Britain was over, and from now on the Allies would be confidently building up their forces to a realistic gaol. In the forces, we trained, drank beer when off-duty and listened to our number one pin-up girl, Vera Lynn, who will be remembered long after other mundane details are forgotten.

Sunderland I remember. My first posting was as officer in charge of two detachments, each manning an Oerlikan 20 mm Light ack-ack gun, on a bridge in Sunderland, Yorkshire. ADGB (Air defense of Great Britain) sentries would be on duty for two hours on, four hours off - a monotonous ritual, varied only by the occasional forty-eight hours leave. Shortly after I took over, however, the monotony was enlivened - not by enemy aircraft - but by a sentry firing one of the guns by accident. Although this might seem to the reader like a trivial incident, to my Battery Commander it was a serious breach of regulations; moreover a court of enquiry was necessary (a) to account for the expended ammunition, and (b) to pass on the appropriate "rocket" to the officer responsible - me.

But worse was to come. A few days later, the guns were removed, for use by the Navy, leaving us on the bridge like Horatius, with just ourselves and our small arms. And even worse than that again, the "Brigadier", in whose presence all lesser ranks stood palpitating, was due to inspect us imminently.

Yet all was not lost. I have to admit that there is nothing like a "rocket" to stimulate ones efforts to excel. By the time the Brigadier came, I am happy to say, we had transformed the bridge into a bristling fortress - with the help of two tripods holding broom handles (in facsimile of the Oerlikans), and a gallon of whitewash which my No 1 had thoughtfully scrounged. We whitewashed with gay abandon, and were duly complimented by the Brigadier, and my Battery Commander (a confirmation that my stock had risen again).

Middleton I remember. This was where I had my next assignment. From this Air-Force station in Yorkshire, Halifax bombers went on nightly raids to Germany. We had four sections of Lewis guns (certainly an improvement over no guns at all) deployed in slit trenches around the perimeter of the air-field. As students of military history know, the Lewis gun was a unique weapon. It had been around since 1911, and its most notable characteristic was a tendency to suddenly stop firing. There were, in fact, over twenty-two ways in which this weapon could jam; moreover the process of fixing it was complicated by the fact that several parts looked similar, and though interchangeable, would not function if exchanged. It was also possible to reassemble the gun with some parts put in backwards, after which it definitely would not fire. We had been given a nodding aquaintance with this gun, but I didn't really expect to come across it.

Scarborough I also remember briefly, whence, by dint of applying for overseas service (the news of Allied landings in North Africa had just been released), I was posted to Woolwich Arsenal. From Woolwich, I went by troopship to Algiers, and thence to an artillery reinforcement depot at Cap Matifou.

OVERSEAS.

"Yes; quaint and curious war is!"

– Thomas Hardy (1840–1928)

The Casbah was not a salubrious area, and the exterior of the Sphinx, with its massive, iron-clad door, grille and peep-hole was forbidding. However, once inside, there were lights and comfortable furnishings and a well-stocked bar, to which the officers one and all migrated.

The reason we found ourselves in these supposedly exotic surroundings was due to excessive boredom. Cap Matifou, where the ARTD (artillery reinforcement training depot) was situated, was twenty miles from Algiers along a cliff-girt coast. In my memory it was synonymous with limbo. Algiers, on the other hand, which implied the bar of the Aletti Hotel, was more interesting. So, when two majors went out of their way to suggest a visit to the Casbah, they got a good response.

The Sphinx, which we intended to inspect unofficially, was a large brothel, set up by the French army, and located in the Casbah (which was ostensibly off-limits). Actually, we found the bar at the Sphinx to be hardly different from the bar at the Aletti Hotel which we had just left; and our group would probably have stayed there all evening, impervious to the somewhat unalluring charms of the few pathetic girls who were hanging around.

However, at ten o'clock, a well-built lady, with lungs of brass, whom I took to be one of the management, took a stand at the top of the stairs and declaimed in resounding French:-

"Exhibition! Exhibition!"

This was what we had come to see, and the bar emptied; the soldiery carrying their drinks downstairs. Where, soon enough, in pranced two nude young females, one of them with a large rubber phallus (which must have been at least a foot long), which she brandished on top of her head.

"Eh, Voila! Le coq!"

She then proceeded to strap the thing around her waist, where it hung grotesquely at halfmast - a most ineffective parody of an erection. Then the two of them went through a few gymnastics, without too much enthusiasm, and it occurred to me that after having had a long hard day, they were probably tired. However, they were spurred on by the cheers of the audience, who were determined to get their money's worth, and who did everything but pat them on the bottom during the action:

"Hey, comme le chien! - how about comme le chien!"
"Soixante-neuf! Come on, give us the soixante neuf!"

It was pretty obvious that the girls weren't too keen on the soixante-neuf position, but they dutifully demonstrated it, for all of two seconds, after which - the piece-de-resistance - the second girl whipped a cigarette from the mouth of one of the audience, deftly inserted it in her crotch, then blew the smoke out of her mouth. Much applause! Then, with a flourish, she gave the cigarette back to its owner, who promptly trod on it.

So *this* was the Casbah - not a bit like it was portrayed in the Charles Boyer movie . . . . . .

From now on my army career was to resemble a travelogue. No conducted tour could have included so many different and varied changes of scene.

My next posting was to a Light Ack-Ack Regiment defending installations in the Line of Communications. From then on, dogsbody of a second lieutenant though I might be, I had a job, and at times could even believe that I was doing something useful. Indeed, from this time on, it seemed that the war had become a more serious matter, for our unit became part of the "Beach Group" in support of the landing of the second Canadian Division at Pachino, Sicily. Our function was to land behind the infantry, deploy our guns around the perimeter of the beach-head, and stay there as long as the beach was used for landing supplies. In the event this was a fortunate landing, in that it was unopposed, and we would stay on the beach until the ports of Syracuse and Augusta had been captured.

We moved up the North African coast behind the First Army into Tunisia. Sousse I remember only for its clouds of dust. Sfax I remember because that was where we embarked for Sicily, on a night when gale force winds were sweeping the Mediterranean. I remember the voyage too, where we could look ahead to a line of landing ships, all like our own being top-heavy with trucks, tanks and equipment, and rolling their bows under and their screws out of the water in the heavy seas. On either beam were destroyers of the Royal Navy. It was an impressive armada - impressive enough to keep me, surprisingly, from being sea-sick, although most of our gunners were prostrate.

The beach-head at Pachino, I also remember. Fortunately it was unopposed, as it took an inordinately long time to unload our guns in the shallow water before we deployed them. We shared the beach with the barbed-wire and mines, and stayed there for far longer than I expected. From the beach-head, we moved up behind the Eighth Army, our role being part of the defensive barrages around the ports of Syracuse, Augusta and Katania. Then we followed across to Reggio, and up the toe of the Italian peninsula. At Reggio I caught infectious hepatitis, a common disease in that theatre of war, second only to malaria, spent two weeks in hospital, and two weeks at a Red Cross convalescent home in Taormina. Here I had time to admire the ancient Greek theatre, and fish for squid in the blue waters off shore.

Bari, I also remember. Our guns were part of the barrage defending that port during the disastrous German air-raid of December 2nd 1943. Over twenty ships in the harbor were sunk, including tankers and an ammunition ship, which inevitably blew up, causing tremendous damage. There were over a thousand casualties. The official account of this raid was played down, but we knew what a setback it was.

A more impartial account of this happening perforce had to wait until Glenn B. Infield's book "Disaster at Bari" was published by MacMillan in 1971*. In this report, the carnage at the docks to which we were all eye-witness was well-documented; and there were other more sinister overtones referring to stocks of chemical weapons held by the Allies, which made for very revealing reading.

From Bari we moved to Foggia plain, where we deployed our guns around the airfield. From here, Flying Fortresses flew to Britain, bombing Germany on the way, while an equal number of bombers left Britain, bombed Germany and landed in Foggia. At this point my natural impulsiveness again got the better of me (boredom had once again set in), and I volunteered for the infantry. I felt that I could disregard the axiomabout never volunteering, as I had volunteered several times and it had done me no harm. So a short while later I found myself en route to Avellino, near Naples.

In any case, due to the declining activity of the Luftwaffe, our anti-aircraft role had become redundant, and a few months later the whole regiment to which I had recently belonged was drafted into the infantry without option.

AND STILL MORE....

"Soldiers who wish to be a hero
Are practically zero;
But those who wish to be civilians,
Jesus, they run into millions!"
Army latrine inscription -

– quoted by Norman Rosten.

It was spring 1944. The Allies had just broken out of the Anzio beachhead, and Rome was about to fall. At Avellino we took junior leader's courses, and a mountain warfare course.

"Taffy" Harris came from South Wales. An ex-engineering student at the start of the war, he had been a miner, had played rugby for Neath, and was a welter-weight boxer of no mean ability. After the mountain warfare course, four of us spent a month at the S.A.S. training depot. This group, under Col. Stirling, were dropping paratroops and supplies into Albania and Yugoslavia, to aid the partisans. I wasn't too sure how I would react to jumping out of a plane, but I figured that if Taffy could do it, so could I. As it happened we did not get the chance. We spent a very instructive month being trained in guerilla tactics on the shores of the Adriatic. The training was remniscent of being in the Commando (the S.A.S. were a "special service" unit), and included riding, packing and looking after mules. We were extremely fit physically at the end of it.

Taffy was a charming man, with a marked sense of humour. Some Welshmen might not admit to the fact of their ancestors having painted themselves with woad, but Taffy was proud of it. He was a very good friend. We went on several occasions to Naples, staying at the home of a charming (and moral) lady known as "the baroness", who always had a room available. We never got into serious trouble in our forays to Naples (although that was not Taffy's fault), but we did have a lot of fun. In fact I must admit that towards the end of our comradeship, I sometimes found it prudent to leave the party after a certain stage had been reached.

There were, in fact, three stages in Taffy's behaviour, dependant on the amount of liquor consumed. The first was simply good humour. When the second stage was reached, Taffy wanted to sing - he had two favorite songs, both of which, being a Welshman and a tenor, he sang well: the Rose of Tralee was one, the other was Cwm Rhondda. He needed no encouragement to sing. However, the second stage often merged into the third, and in the third stage Taffy wanted to fight. He wasn't at all particular as to whom he fought. Only once did he pick on me (he apologised later), but it was obvious that the identity of the target was not important, only that there *must be* a target. He was a formidable opponent, and the change from stage two (happy) to stage three (mean) was startling. If there was no immediate prospect of a fight, Taffy would proceed to start one. One sure-fire way was to select a likely-looking person on the street, push him violently, then say: "Why the hell don't you look where you're going?" Usually Taffy was sober enough to pick on an officer of the same rank as himself, and very often the victims friends would also join in, and a good time would be had by all. But one time, having drunk more than usual, he accosted a Brigadier, red tabs and all. Needless to say, Taffy had the worst of that encounter, but I feel sure that he cherished a secret hope that one day a more senior officer would strike him.

There were other interesting places to visit besides Naples - Pompeii was worth a visit, and we even spent forty-eight hours on the island of Ischia. But it was not necessary to go so far afield. Taffy and I made friends with two Italians, daughters of the local schoolteacher, and many evenings were spent - well chaperoned - at the manse; a very pleasant way to learn Italian.

Soon the postings came, mine for the Ist Bn Loyals. Taffy hoped to go to a South Wales regiment. I often wondered what became of him. The S.A.S certainly made a mistake in rejecting a man of such tremendous stamina.

The first battalion of the Loyals (the Loyal North Lancashire Regiment) were part of the First British Infantry Division. They had recently been through the ordeal of the Anzio landings, and defence of that beach-head. Since breaking out from there they had been continually in action, advancing with the Division past Rome and Florence. The C.O. was a very competent soldier, respected by all. The other ranks were nearly all from the north of England, forthright and genuine, with a few Londoners. There was a sense of confidence and purpose in its well-integrated ranks. I was eventually called up when the battalion took over positions overlooking Bologna, on a mountain salient called Monte Grande. We would normally spend ten days in the line, and move to the rear for a five day rest.

I was given a platoon of B Company, which occupied a spur running out from the summit of the mountain. What made this position interesting was that on the actual summit was a trig. point, with a large survey signal still erected above it, providing a perfect ranging point for artillery, marked on every map and completely visible; and my dugout was directly beneath it.

We lived in such dug-outs, holes in the ground with sand-bagged roofs. The roofs, although constructed with that useful piece of waterproof issue, the gas-cape, leaked. We were wet for the whole ten days and we slept with our boots on. I myself got very little sleep at all, as it was part of my job to see that all sentries were awake at night. But apart from that chore, it would have been very difficult to sleep anyway, as Jerry fired everything at us that was available for twenty-four hours of the day, including shells from the six-barreled mortars (we called them "moaning minnies" - fortunately their aim wasn't very accurate).

But the seventy-eight mm mortar bombs *were* accurate. There was four inches of snow when we arrived, and we wore snow-suits for camouflage. These were white cotton overalls. Unfortunately, after a few hours crouched in a dug-out, they changed colour, and were not so useful.

Should one wish to "go to the bathroom", the usual way would be to have all buttons undone, toilet paper in hand, and wait for a lull in the rain of hardware that was landing on the hill. Then, when conditions quieted somewhat, one would make a quick dash of about forty feet, to a shell-crater, situated conveniently near the dugout. If during the ensuing proceedings, one heard a sound like: "Thup!", it meant that a German mortar had just gone off, probably aimed at oneself, and that meant that one had about twenty seconds to get back in ones hole. I almost "bought it" on one such occasion, landing head and shoulders in the dug-out, while the mortar shell landed simultaneously five feet away, fortunately in a hollow.

The second time that we occupied the site, it was much quieter. One could even move - with alacrity - in the daytime, without feeling that one was in somebody's gunsights. The third time was dead quiet, forbodingly so, and when time came for our relief (by the Americans who, unaware of the possible consequences, stood around in droves on the hilltop - hulking figures in the gloom of the night, wearing greatcoats with blanket rolls strapped aroung their huge packs) - we were glad to get out of there.

Each infantry division in Italy had been withdrawn, in rotation, to spend a period of "rest" in Palestine. Our turn came early in 1945, and after embarking at Taranto and landing at Haifa, we spent the next few months in various permanent army camps in that pleasant countryside, as it was then, reorganizing, getting back to full strength, training and catching up on quotas of leave. Competitive sports assumed a position of importance in day to day events. We played soccer, held swimming meets and inter-regimental boxing matches. We even played water-polo. On leave we saw Cairo, the Pyramids and Shepherd's Hotel as thousands of troops before had done.

On one instructive occasion, we went on a tour of Jerusalem conducted by a former history professor, now a sergeant in the Education

Corps. With him as guide, the past came to life, as we visited the Stations of the Cross, and other traditional landmarks. Yet, no matter how devout a Christian - and there were several in the party - one could hardly be less than cynical after seeing the small bits of wood being sold as "pieces of the true cross", and after examining the "footprint" on the Mount of Olives, and after seeing the immense wealth in gold, silver and precious stones made into altar vessels in the Church of the Holy Sepulchre. The sergeant explained that many of the landmarks were quite unauthentic, research having proved the correct locations to be elsewhere; however, the attributed location would obviously never be changed in view of the hundreds of thousands of visitors. The last site we visited was the Mount of Olives, where the site of the Ascension was occupied, not by a church, but by a mosque. The Christian tourist, after leaving his shoes at the entrance, and viewing the hallowed spot, contributes his dollar at the conveniently placed alms box at the exit, for the benefit of Allah.

The war in Europe ended while we were in Palestine. This was for us the end of our "rest" period. From that time onwards, we were involved in endless call-outs to man road-blocks, quell civil disturbances and generally back-up the Palestine Police in their efforts to curb the violence that was erupting on both sides of the Jew versus Arab dispute. The main body of Jewish settlers were law-abiding, but the extremists were anything but so, and the violence culminated with the murder of the British High Commissioner, and soon after that of Count Bernadotte - the Swedish diplomat, well-respected by both sides, who had been appointed by Geneva as mediator; these particular murders being two of the first to be carried out by the Stern Gang.

The basic problems of the Middle East were all present in 1945. Of today's volatile state-of-affairs I know only as much as I hear and see on the media, but I can certainly comment on the situation we faced forty-five years ago. In fact, a little past history might help straighten out the record which has become somewhat muddied by poorly researched novels and other books purporting to describe those times.

In 1945, when we had arrived in the Middle East, a truce for the duration of the war had existed between Jew and Arab. The last "troubles" in Palestine had been in the late 1930's, when the Arabs declared a "holy war", and attempted to overrun the newly formed Jewish state, and drive the Jews into the sea. The British, whose mandate was to administer the state, had defended the Jews, and when the Arab revolt was over, had built strong reinforced-concrete police stations - block houses - in case such an event was repeated. As an additional defence against possible Arab attack, each kibbutz was allowed an arms cache, and members of the kibbutz were allowed to train militarily for self-defence. This was the "Haganah" - a legal and purely defensive organization - *not* some kind of underground terrorist group, as I have heard it described by journalists who prefer sensationalism to fact. I have to say that when I read some of the supposedly researched accounts of those days, I sometimes wonder if they are describing the same country where we spent time endeavouring to keep the peace.

Unfortunately for the British in 1945, all their fortified police stations were in the predominantly Arab parts of Palestine; and in 1945-6, it was not the Arabs who fomented civil riots, but the Jews. It was a politically impossible situation. Quotas for immigration to Israel set by the League of Nations did not begin to cover the thousands of homeless Jews in Europe, who were determined to enter the country; while violence from the Arabs waited to erupt if the quotas were exceeded. The violence was chiefly from Jewish extremists who were seldom caught.

Our job would often be to surround a settlement where a suspected terrorist had gone into hiding, while the Palestine Police went in to interrogate. Often the police used dogs, but although the dog did everything but speak when it recognized the criminal, the police were powerless unless some person would give evidence in court. This seldom happened, as kibbutz members would never turn in one of their own people; moreover, if anyone did agree to do so, he would in all probability be found murdered before he could testify.

As soon as "VE" day was announced, our battalion was ordered up to Syria. Here, while the war had been going on, the Vichy French government had been in control. Now the French were about to depart, and since they were leaving a very hostile population behind, we were to see that they left unmolested. In one case we were too late. The Jebel-Druse, who had been formed into several French colonial regiments (having French officers, and Arab NCOs), reacted quickly. They lined up all their officers against the wall of the fort, then turned a machine gun on them. When we got there, all was peaceful; the ring-leaders willingly gave themselves up, knowing tbat they would be tried in a civil court, and having high hopes of being dealt with lightly.

I remember a wealth of detail from that period of my army life, but none of it today seems noteworthy. We were well-drilled puppets. We paraded, held guard mounting ceremonies, took courses, drove around the country in convoy, wrote letters that couldn't mention any news, and for entertainment had mess parties, when neighbouring units were invited over, and much horseplay resulted. We wore bush jackets and shorts, and often desert boots (known as brothel-creepers), carried fly-swatters, and interlaced our every-day speech with pseudo-arabic slang. Army life in the Middle East was a phenomenon which will be remembered, though not necessarily with nostalgia.

When it became apparent that the end of the war was inevitable, the Powers That Be introduced several new ideas. There were current affairs lectures, and debates, and numerous publications for the soldier returning to "civvy street". One such paper that I read, and two books that I picked up in Cairo when on leave, seem to have determined my future profession. The first book was Searle's Field Engineering, the second - Chambers Seven Figure Mathematical Tables. I lost no time in writing to my uncle in British Columbia, Canada. The Powers That Be also issued us with a "demob number", based on an aggregate of age and length of service, mine being 24.

In the spring of 1946, in company with two brother officers - also time-expired - I proceeded to the depot at Sidi-Bishr in the canal zone, en route for "blighty"; where in due course we were to hand in our pay books, and receive in return a new suit, and a small gratuity as a parting gift from the army that had employed us for the past six and a half years.

4. BRITISH COLUMBIA

WESTBOUND LIMEY.

"Come, my friends, 'tis not too late to seek a newer world".

– *Alfred Lord Tennyson (1809–1892)*

I had no preconceived ideas about Canada, apart from the notion that it was a friendly country, whose inhabitants spoke English and/or French, probably wore snowshoes in the winter (except in Vancouver, where I was bound), and was part of the Commonwealth. The word "colonial" had yet to acquire a sinister meaning for me, which suggests a certain amount of naivete on my part.

However, after a visit to British Columbia House, and Canada House in London, England, I found out that, although Canada was part of the Commonwealth, one would still require a visa in order to get there. To get the visa it was necessary to have (a) a passport, and (b) a steamship ticket, booked for a certain date. The first I had. The second posed what seemed to be an unsurmountable problem, since this was June 1946, and only nine months after the end of the war.

I visited the offices of the Cunard Shipping Company in London. The counter was at least fifty feet long. There was one bored clerk sitting behind it, reading a newspaper. There were no other customers, only myself. It was apparent that either there were no ships, or no berths to be sold. It turned out to be the latter.

The clerk was very helpful. He explained that the British government had commandeered all berths at the beginning of the war, and now issued tickets to travellers according to a system of priorities. Therefore, I should apply to the proper government department, on the appropriate form, and obtain a priority. My passage would then be attended to as a matter of routine. He added:

"Of course - if you had a visa, I could sell you a ticket right away, as, due to last minute cancellations, the ships all sail with a few empty berths".

So it seemed that without a steamship ticket - no visa; and without a visa - no steamship ticket. And since my priority was so close to the bottom of the list, I would have to wait about three years. I saw myself as a potential victim of bureaucracy. Yet I had, for the past six and a half years, been part of a very large bureaucracy myself, that of the armed forces; and one of the things which one learns in such an establishment is that there are ways and means of thwarting the due processes, if one can only discover them.

At this particular juncture, it occurred to me to enquire what the waiting period would be for a *flight* to North America. To my great surprise, I was able to book a passage by air to New York, thanks to British Export Airlines (a forerunner of BOAC) for a date two months ahead. I could not wait to get this in writing. No deposit was required (this *must* have been an oversight).

From that time on, all doors seemed to open to me. First those of Canada House, where on the strength of my airline booking and vaccination certificate, I was able to obtain a visa. Then I proceeded to the United States Consulate, where upon production of my Canadian visa, I was given a U.S. visa. Then to the Cunard Shipping Company offices in Liverpool, with my visas and my money in my little eager hand, where the friendly shipping clerk said:

"Too bad, you just missed today's sailing to Halifax, but the S.S.Gripsholm sails tomorrow, and we can give you a berth on her".

So on a day in June 1946, after being for six weeks a civilian, I finally started to travel westwards, landing in New York a week later; after which

I wrote to the airline company cancelling my proposed flight (which I could not have afforded in any case).

The S.S.Gripsholm had been a hospital ship during the war, and this voyage was its first as a passenger liner since hostilities had ended.

I shared a cabin with three other Britishers. One was a breeder of fox-terrier dogs, on the way to a dog show in New York. We didn't see the dogs, but they must have been on board somewhere. I wondered if they, too, had priorities from the appropriate government department.

I shared a table with another Briton and his wife, who were emigrating to the U.S., and with a returning Canadian army captain, who was in uniform. It occurred to me that I would be more comfortable dressed in this fashion, and I took to wearing mine, for the last time. I really felt more at home in it than in the ill-fitting suit, a parting gift of the British government, which was my only presentable item of clothing. It is hard to visualize now, but at that time the majority of people in Britain wore a uniform of one kind or another; and the same proportion seemed to hold on board the S.S.Gripsholm.

However, that changed when we arrived in New York. Here there were few uniforms, and it seemed a long way from war-weary Europe. During the voyage I had made friends with a very attractive girl who, with her mother, was travelling to stay with relatives in New York. So it seemed logical to spend the next week in the big city, getting acclimatized to the New World; and a very good time we had, too, "doing" the sights that all visitors like to see - Statue of Liberty, Radio City, Carnegie Hall, Central Park (where, in 1946, it was safe to go for a walk); we even went to Coney Island one afternoon.

But after a week, I felt the urge to continue my journey west. I also felt the need to conserve my dwindling finances - New York *was* expensive, even if I was staying at the Y.M.C.A.

So one day I packed up, said goodbye to the girl-friend, and climbed aboard a train to Montreal. There I bought a one-way ticket to Vancouver, and travelled all the way in the day-coach. Being young, I thought I could economize on sleep. I had sent a wire to my aunt to tell her that I was coming, and bless her, she was there at the C.P.R. station in Vancouver to meet me.

VANCOUVER.

"... an ideal home for the human race, not too cold, not too hot, not too wet, and not too dry, except in hotels - a thing which time may remedy".

– Stephen Leacock (1869–1944).

A friend of mine recently recently described Vancouver as it was in the late 1940's. He said:- "You know, a dog could have laid down in the middle of Broadway and Granville, and gone to sleep, and he wouldn't have had any trouble".

This was gross exaggeration, as my friend well knew. The street-cars which used to cross that intersection every two minutes would have reduced that unfortunate dog to strawberry jam (as the poet said), in short order. However, there are a multitude of people who remember Vancouver from those days, and who would agree with my friend that it was a much more relaxed city then, than it is now.

My first impressions were in line with Stephen Leacock's assessment. I could go further - I found it neither too small nor too large, but entirely agreeable. Many old landmarks were in place in 1946, e.g. The old Vancouver Hotel - old Granville Street Bridge - old Georgia Viaduct - B.C.Electric terminals at Carrall Street, and at 25th and Main - the RCMP "horse-barns" at 35th and Heather - Union Steamship wharf - to mention but a few. All of these are now gone, though some lasted longer than others. The only tall buildings were the Marine Building, the Sun

Tower, and the Sylvia Hotel ("dine in the sky") - which are still with us. But there was no B.C.Hydro building, no Empire Stadium, no Empire Pool, no new Granville St. Bridge or Georgia Viaduct, nor B.C.Place and Expo site. For that matter there was no Royal Centre, or Pacific Centre, or any of the myriad shopping malls above and below street level, nor any buildings higher than five stories in the West End. The Canucks were there alright, but they were in the Western League, and they played at the Forum. One paid from 50 cents to $2 to go and jeer at them. Granville Island was a purely industrial area, as was False Creek, and noone would have believed that a fashionable market-place would locate on the former site, while condominiums would flourish on the latter. Nor would they have envisioned the transformation of part of Skid-row into the present Gastown.

Getting from one part of Vancouver to another at that time was easy. Every main street had its street-car tracks, and the routes were easy to memorize. My uncle and aunt had an apartment in Mount Pleasant, near 7th and Carolina, convenient to the Circle street-car route which ran down Main Street, along Hastings, up Granville and back via Broadway - or one could go the opposite way. That route went right into the heart of the city, business offices, C.P.R., downtown hotels, Woodwards, Spencers (prior to Eatons, and now prior to Sears), and the Bay, although then it was called the Hudson's Bay Company. Also, the Imperial Bank of Canada, and the Canadian Bank of Commerce were two separate institutions, at either of which one could obtain Canada Savings Bonds with an interest coupon of - believe it or not - two and a half per cent.

Cars were not numerous. No new cars had been manufactured for civilian use for several years, since the USA had entered the war. On the other hand, the traveling public could hardly complain, as, for a quarter they could buy six street-car tickets, and get to any part of the city by transfer. Five years later the street-cars were replaced by buses; the last to go being the (in)famous Oak Street Car.

My uncle was a tug-boat skipper. He was called "Cap" by most of his friends, and he was away from home for a week or so at a time. Most of the time he was towing log-booms, or Davis rafts down to Vancouver from camps and ports on the B.C. Coast. When back in town, he usually had several days off. My aunt had a short wave band on her radio, and was able to tune in the conversations from the tugboat's radio-phone, so she usually knew when he was due back. He was a very dynamic man, and youthful in his ways - dressed "young", and acted the same, although he was in his late fifties.

He was in fact due back a few days after I arrived, and from their somewhat unconventional greeting (probably for my benefit) - it was apparent that both took a good-humored view of life.

'Hello, you big bum'! said she.

'How are you, you old bag'! he replied.

After which, if nothing of importance were to happen, he would arrange to meet some of his cronies at the Arctic Club before dinner.

This was a very necessary institution at that time, as, although one could buy a beer in a beer-parlor, or a whole bottle of hard liquor at the government liquor store, nowhere was it legal to buy liquor by the glass; except at a private club such as the Arctic Club or the Pacific Athletic Club. I seem to remember that if one went to a dance, or a night club, one brought one's bottle and put it under the table, while the waiter produced glasses and mixers for the occasion. It was a pretty barbaric system, and one which the government seemed reluctant to change. The beer in the beer-parlors was served two glasses at a time, and the objective seemed to be to fill a person up until he couldn't drink any more, and then to push him out on to the street. Hotels were all dry. There were no cocktail bars, no wine list with meals; so one took one's bottle up to the hotel room, and sent for the ice and the mixer; and if one ran out of booze after six pm, one called a cab driver, who would get another bottle from a bootlegger.

My aunt was a very different kind of person from my uncle. Together with three other sisters, she had come to British Columbia from Aberdeen when she was a girl, and she still had a slightly Scots accent.

She was a very good-hearted and good-humored lady, although at times she could be as dour as they make them in that part of Scotland.

My uncle had for many years been the starter for the swimming events at the Kelowna Regatta. One of his personal friends was a photographer on the staff of the Vancouver Sun. Just prior to the start of the 1947 Regatta, a full page picture appeared on page one of the Sports section of the newspaper. It showed my uncle, beside the pool, with his eyes crossed, holding the starting pistol to his temple. I forget the caption (there was a lot of the ham-actor in my uncle's make up) and it certainly was a convincing pose.

I don't remember anyone who didn't find the picture hilarious, or at least highly entertaining - except my aunt. Her reaction was quite firm:

"The very idea! Making such an exhibition of himself!"

* * *

Vancouver 1946. . . . .

Dal Richards was there, and Lorraine McAllister. They played at the Commodore, or at the Roof in the new Vancouver Hotel - or at the Cave. Remember Spike Jones ('Hello Music Lovers')?

The old Vancouver Hotel stood at the south-west corner of Georgia and Granville. It was taken over by returning veterans that year (I forget the details), and I remember going there to visit a friend. I also remember, a few years later, peering down into the huge excavation after it was demolished to make way for Eatons and then Sears. Birks Clock - alas now gone to a different location - stood at the South East corner of that intersection, and it was there that you met your girl-friend before taking her out to dinner. Not that there was a great choice of restaurants as there came to be later-on.

I remember Vancouver in 1946 because it marked a turning point in my life. In that year I became a B.C.resident, and later a Canadian citizen, an ex-Brit if you will, and started a new career.

1946 was also a memorable year for Vancouver - a turning point from which it never looked back. It was off with the old, and on with the new.

Now one sees a city of mushrooming high-rises and gleaming shopping malls, and industry has been banished to the outskirts of town. It is still the same city. But for this particular resident it is Stanley Park, Spanish Banks, the Lion's Gate and North Shore mountains, and miles and miles of tree-lined boulevards which induce nostalgia - and of course those parts of the old city which exist now only in memory.

WOODS PRELUDE.

*"I wish to preach, not the doctrine of ignoble ease,
but the doctrine of the strenuous life".*

– *Theodore Roosevelt (1858–1919)*

My career as a logger was fairly brief. The weak mind I had, but not the strong back. But I did relish the experience, and everything I learned was new.

In the British Columbia woods one needs a pair of caulked boots. So I was outfitted at Paris', in downtown Vancouver, where I had a pair made to measure. This was the first time I ever had boots made to measure, but I was assured that it was quite customary. Then to Edwards Lipsett's, where I bought "Dryback" pants and jacket (they weren't), mackinaw, waterproof hat and lunch bucket. I don't remember ever wearing a hard-hat at the time; I only saw fallers wearing them. Then I was issued with a Union Steamships Co. voucher, and several meal tickets, and climbed aboard a craft called the S.S.Venture, which steamed off up the coast, and took two days to reach Loughborough Inlet. It was quite full of large, bleary-eyed men, with packsacks and bedrolls. Most were hung-over, and only wanted to get into their bunks and sleep. The sleeping accommodation reminded me of a troop ship; but I do not remember much else about that voyage. We must have stopped at isolated wharfs about half-a-dozen times. It was all a new experience for me.

I met two or three other men going to the same camp, and upon arrival we were assigned to huts. I was in a hut with about eight beds. All the other occupants of the hut were about my age, but all were experienced in the woods except one, who was a returned Canadian veteran. This man, who had been hired as a chokerman, quit after a month, so I was the only "green" one left. I do not remember the names of any of the crew in the hut, but all were eminently congenial people. One drove the logging truck, another was the chaser, another the rigging-slinger and several alternated at various different jobs.

From the start I felt at a disadvantage. For, while every red-blooded Canadian boy had learned how to use an axe since infancy, I had not. Not only did one need to know how to use an axe, but also how to pull one end of a cross-cut saw. This is the tool that separates the men from the boys, and I was obviously still a boy. However, after two months I did manage to show some proficiency, and acquired a whole new set of muscles in the process. It is interesting to note that now, in the year 2011, few Canadian boys, red-blooded or otherwise know how to use an axe; a power-saw maybe, but not an axe.

It must have been quite a novelty for the others to have a real, live Limey in the camp. There was a certain amount of good-natured kidding, of course, but nothing that I couldn't weather. The daily routine was high pressured. There were no coffee breaks. Saturday was a full working day. Only Sunday was free, and it was spent by the wise ones in resting on ones bunk. Some other wise ones, and some not so wise, played poker. In our hut we played penny-ante, but there were some fairly big games in other huts, and a lot of money changed hands.

In the mess-hall, when the "gut-hammer" sounded, about thirty men sat down wordlessly, and commenced to eat. The only conversation would be "Pass the salt!", or "Pass the jam!". I was told that this was the custom, so that the meal could be quickly finished, and the cook and flunkey could then get on with cleaning up the tables afterwards. But the food! For six and a half years of war and food rationing, I had not seen such variety, and there was an unending supply of everything. The flunkey would bring up a platter with six T-bone steaks on it, but before the platter touched down on the table, the six steaks would be gone - speared by six forks, wielded by six hungry men; and the flunkey went back for more. So the meal went on, having started with soup, through fresh salmon, beefsteak, and lemon meringue pie. There was coffee afterwards - if you had room for it - and then, some of the men would be making up a snack and thermos bottle to take back to the hut, in case they got hungry before going to bed.

It was quite a surprise when one day I mentioned to one of the men how good the food was. He didn't agree with me:

"The food's not that good ", he said, "Most of the guys are complaining about it".

And he must have been right, because a few days later, the cook was fired, and a new one arrived on the boat (*"Now* we'll get some good grub for a change!") As for me, I didn't notice much difference - I was too busy overeating in order to make up for all the energy expended during the day.

For the first day, our crew consisted of the engineer, the bull-bucker and myself, after which I was judged fit to be a surveyor's helper on my own. We ran lines for logging roads, blazed out timber limits, and then laid out settings for the logging operations.

After a while the engineer left, so the superintendent put me on the bull-gang. I well understood why it was called that, but not why he put me on it. This was the crew that laid planks for the corduroy logging road. Hemlock, at that time, was not usually cut for lumber. But it made excellent two-inch planks, and was more available than gravel or crushed rock. The gang foreman was a Swede (another one) - a very popular man, but also a driver. The work was done literally at the double; planks were unloaded, and carried at the trot, spiked down, and then you trotted back for more.

I engaged in a number of different activities while at this camp. I had a spell of setting chokers, worked in the mill, again with the engineer, and finally - the back-rigging.

In 1946, high-lead logging had not changed much from when it was first introduced in the early 1900's. I understand that the principles are no different now, although today's massive equipment has eliminated many of the earlier practices. In those days, a large Douglas fir would be cut down, topped, limbed, rigged and then re-erected to be used as a spar-tree. After it was upright and stayed, a big block would be hoisted to the top, and strapped in place. This was the basis for the landing, at which a donkey-engine operated, dragging in logs to the "cold-deck" pile. Hauling in logs was, of course, production. Rigging the spar-tree, however, which took time, was pure overhead, but it was a very necessary part of the operation; so, in our case, one crew was always ahead rigging the next spar-tree, while the other crew logged.

In this particular case that I remember, no road had yet been constructed to the next proposed landing, yet it was essential for the donkey engine to be moved to the landing - through bush and slash - so that a spar tree could be rigged. To my untrained mind, this seemed an unsurmountable problem. However, it was no problem for the rigging crew; in fact, moving the donkey-engine on its sleigh through the woods was obviously standard procedure. There is no doubt that it was a pretty spectacular operation.

At this point perhaps a short explanation of the preocedure is warranted, as I am sure it is many years since loggers have ever had to use it. The first step would be to select a large stump 200 or so feet ahead of the donkey engine. To this stump would be attached blocks, starting with 60 lbs ones (for the straw line), and ending with a huge one (that probably weighed 500 lbs). Now, with the main line cable from the donkey's capstan, through the block and back, with the end fastened to the sleigh, if full power was applied, the donkey itself would perforce have to move also - towards the stump.Full marks for efficiency, and again full marks for the crew who accomplished the seemingly impossible.

Especially the crew, because although the principle sounds easy, the practice was definitely not so, as the manual work was done amid a sea of logging slash and downed trees, pulling straw-line, carrying 60 lbs blocks often balanced on criss-crossed logs 6 feet above the ground, and with a running commentary from the man in charge.

The man in charge of the project in this camp - the hooker - was an unlovely character with a loud mouth and Quebec accent. Hearing this man's instructions was an education in itself, as he possessed a longer string of expletives than I had heard to date (in spite of 6 years in the army). He cursed and blasted both men and materials with impunity. We, the crew, stoically endured and ignored it all, while doing what was necessary on the job.

Perhaps harsh treatment was necessary because of the arduous nature of the job, although later experience tells me that there could have been better ways. Anyway, the jobs got done in spite of his exhortations (not because of them), and the crew's considered opinion of their boss could be summed up in two choice words from his own vocabulary.

There is no doubt that it was a spectactular operation. After we had watched the huge bull-block go thrashing through the bush, pulled out by the donkey-engine, *then*, accompanied by more expletives from the hooker, got the block strapped in place to the stump, and the mainline back to the donkey - then - we could stand back and watch with a sense of achievement as the donkey-engine on its skids, with the hooker standing up front (like an admiral), surged ahead in a series of monstrous jerks over stumps and downed timber for some two hundred feet, the cable vibrating with the strain After which the whole process would be repeated - more straw-line, more sixty pound blocks and so on.

However, after two such sessions, *that* prospect had lost its appeal for me, and I decided that although I relished the experience, it was time to look for some other line of work, preferably one which needed less brawn and more brain. So, when the next Union boat's whistle sounded, I went to the office and got my time.

GLACIAL CREEK.

"Life grants nothing to us mortals without hard work".

– Horace (65–8 BC).

Jervis Inlet is a magnificent fjord. Its blue waters, which are over two thousand feet deep, extend to Queen's Reach and Princess Louisa Inlet, which is half-way through the Coast Range. At that time in 1946, there was a small boat called the "Jervis Express", based at Pender Harbour, which serviced all the camps on the inlet - Goliath Bay, Vancouver Bay, Brittain River, Glacial Creek, Deserted Bay and on to the Malibou Rapids. Not long after coming down from Loughborough Inlet, Dave and I found ourselves on board this craft bound for Glacial Creek, having taken the Union S.S. boat as far as Pender Harbor.

The woods operation of this camp was quite different from the one I had recently left. It was being managed by a firm of forest consultants, who were engaged in preliminary work - mapping, forest inventory and road location, for which I had been hired as a compassman. The timber was old growth - which meant big trees - except for an area where, thirty years before, cedars had been felled and bucked, but not taken out of the woods. At that time logging operations had been suspended; but West Coast cedar is a remarkable wood, and even after all that time had passed, most of the logs were still completely sound, and could be taken out and sold.

Being a novice in the woods, the size of the timber at these camps did not arouse in me any feelings of awe. Fir logs were often forty-eight inches in diameter, and the cedars could well have been sixty. It is hard to reconcile the idea of those trees as we found them, with the later growth that we see in the woods today. Occasionally the disparity is brought home, as for instance when in the sixties I worked in the Vancouver Watershed Reserve, logged in the 1920s by the Capilano Timber Company, and where one could stand in one place amongst the small second growth hemlocks, and count up to half-a-dozen monstrous cedar stumps, none less than ten feet in diameter, and some as much as sixteen.

There are of course many old photographs, and articles which have recorded the massive old-growth forests, but such information has small impact compared with the evidence of ones own eyes. We have become accustomed to seeing smaller trees (any above eight inches diameter can now be logged), so when we get a glimpse of the old-growth in the few places where it can now be seen, such as in the Carmanah valley, or on the West Coast trail, we cannot help but be impressed.

Today we are truly aware of the heritage imposed upon us by such trees, but in the early days of which I write we failed to recognize their uniqueness.

Glacial Creek was a small camp. There was a Superintendent, Bull Bucker, Timekeeper, Cook, Flunkey and Bullcook; also two or three fallers and a small logging crew which fluctuated as required. As well there was a young engineer, and his young (I *was* young) helper. Dave was energetic, pleasant and twenty-four years old, and very fit physically after several seasons of surveying and mapping on the B.C. coast. He was completely outgoing, and if he occasionally used an expression that I hadn't heard before, such as "give her snoose!", then I just did what was necessary, and stored up the phrase for future use myself.

In fact the learning process at that camp was enjoyable, since Dave knew theory as well as practice. I think he found me quite a challenge. My knowledge of trees up to this time had been rudimentary. To me, a tree was simply a tree. However, it wasn't long before I could recognize all the commercially important ones, even by their Latin names. So, if a Douglas Fir wasn't a true fir, then it was without doubt a *pseudo-tsuga taxifolia*. (Although I must record that in every-day language we referred to Amabilis Fir as "balsam", and lodgepole pine as "jack-pine").

The reason why I had applied for a job with this firm, was mainly in order to find out if there might be jobs in the woods other than those in which one set chokers, or pulled straw-line for eight hours a day. Dave, who had recently graduated from U.B.C. was able to reassure me on this score. We were involved in office plotting and calculating as well as field work; and somehow, to spend the day climbing through the salal and underbrush, and come back to camp soaking wet, did not seem such a useless operation if, next day, one could take a scale and protractor, and plot ones progress on a map. And there were days when the rain stopped, the sun shone, the air was clear, and one could smell the forest world. Who would not find such an environment stimulating?

Camp life was always interesting. On a Sunday one could go fishing, or - if not taking part in a "bull-session" - listen to the Sunday radio. The programs were listened to avidly by all - often with six or seven men crowding into our two-man bunkhouse. The radio then was not simply to provide background noise, it was good entertainment. Jack Benny - Blondie - Henry Aldrich - Charlie McCarthy - the Quiz Kids - the Shadow - I could name them all. I even remember the commercials with something like nostalgia - the Philip Morris one, for example, (*"call - for - Philip Morris"*), and the Lucky Strike one ("LS - MFT"). I must say that I have never felt like that about television commercials! In the days of radio only, we must have been less sophisticated.

The Jervis Express went by twice a week in the inlet, bringing our mail, and on several occasions we saw a spanking white ex-corvette go sailing majestically by, complete with cook in tall chef's hat leaning over the rail. This was the ship which - it was rumoured - brought the Hollywood starlets to the Malibou Club at the head of Princess Louisa Inlet. A glimpse of a different world.

My job at Glacial Creek only lasted for about three months. Two weeks before Christmas 1946 the snow came, the camp closed and it was back to Vancouver again.

BUNKHOUSE INTERMEZZO.

"Our horizon is never quite at our elbows".

– Henry David Thoreau (1817–1862)

There would be nothing remartkable these days about a train ride from North Vancouver to Quesnel and beyond, because the route is now just a branch line in the vast network of Canadian National Railways. But let us not forget that these tracks were once known as BC Rail, and before that the Pacific Great Eastern, and its Budd cars carried passengers through a very historic part of B.C.

In the '50's it was known as the P.G.E. (incorporated 1912), and in spite of quite a chequered existance (like a lot of projects initiated by government), it survived till the end of the century; by which time (so I am told) it had even become a profitable institution . . .

In February 1947, however, when I first rode on it, it was certainly not a profitable one, but It was definitely a unique enterprise; and if it is possible for a railway to be known as romantic, then maybe this one would qualify.

As I remembered it then, the P.G.E. was just what it's nickname implied - the "please-go-easy" railway. A great many of the settlements and lodges along the right-of-way had no road access, and were only accessible by rail, so the railway performed a very necessary service.

Many of the land marks for which the stations were named are now gone, but the names themselves continue as reminders of the past.

Some stops were actually named to commemorate historical events of bygone years - such as Birken and Seton Portage (the former after the troopship "Birkenhead" which went down off the African coast in 1852, the latter after Major Seaton, the hero of the day); others to celebrate contemporary history, such as Mons and Marne (after the World War I battles); some were called after, and so perpetuated the names of original settlers; and at least one from a jocular aspect - a whistle-rtop whimsically named "No 10 Downing St", with its neat dwelling, fenced garden and Union Jack flying aloft; to say nothing of the Van-West Logging Company's old station, still known to local residents as "Function Junction". It was a very pleasant railway - for instance, at the flag-stop later named "Currie's", the train *always* stopped for five minutes, usually not for passengers, but so that the train crew could have a cup of coffee, and a piece of Mrs Currie's home-made peach pie.

Stories about the P.G.E. in those early days abound. The most recent one I came across concerned a passenger who asked the comductor what time the train was due to arrive at the terminus; to which the conductor replied: 'What time would you like it to arrive. ?'. .

When it was first constructed, the railway ran as far as Horseshoe Bay in West Vancouver, where it stopped; then started again at Squamish and ran on to Quesnel. In the days before the Lion's Gate Bridge, it carried picknickers and commuters along the North Shore, and to Ambleside, Dunderave, West Bay, Caulfields and Horseshoe Bay, where summer cottages abounded. After the first railway company went out of business, however, the tracks through North and West Vancouver were removed (although the right-of-way was retained), and it became a railway that started in Squamish and ended in Quesnel. Those who had a poor opinion of these two towns used to say that it started nowhere, and ended nowhere.

Such was the P.G.E.Railway in early February 1947, when I disembarked from the Union Steamships boat at the Squamish terminal. The carriages were elegant, with old-fashioned furnishings, and the one in which I travelled could well have been one of the originals. There was a large pot-bellied stove at each end of the coach. One had the choice of sitting in the middle - and freezing - or sitting at the end (and being roasted). I remember poking my head out of the coach when the train stopped at Garibaldi Station, shivering and consciously moving my seat closer to the end. It was becoming noticeably colder than in mild Vancouver.

The camp for which I was bound was at Marne, on the shore of Anderson Lake. It had several bunkhouses, a cookhouse and huts with toilets and showers. It housed the transit-man, two helpers, several drillers, steel men, general labourers, and the usual timekeeper, cook, flunkey and bull-cook. There was also a large, though docile, Doberman Pinscher, who belonged to the timekeeper. The latter was an American, whose name rhymed with "bugle", which was what we called him (he was very fond of blowing his own horn).

One new aspect of this camp was that the cook was definitely "gay". Personally, I found him to be a pretty nice guy; I was of the opinion that if a person wanted to be "gay", that was his business, and none of mine. However, it transpired that the authorities in the company knew all about this, and after a while he got fired - and not for being a bad cook, which seemed rather harsh treatment. After all, he would have been fired anyway as a matter of course, after a couple of months (such periodic firings of the cook being considered necessary to ensure that the camp maintained a good table).

Work at this camp was not arduous. As survey crew our function was to supervise the layout for about six of the steel towers of the first Bridge River Power line. Steel for the towers had not yet arrived; however, due to the current labour shortage, it was considered expedient to keep the construction crews on the job, even if there was no actual constructing to be done. So we did a lot of "filling in" of time. We would walk the tracks to the job site, then back via the very old pack-trail part way up the hill. After work on a friday, we would walk three miles down the tracks to McGillivray Falls where was the nearest post office. Sundry other people who lived thereabouts - settlers and owners of small ranches - would also congregate at McGillivray Falls on mail day, some on foot with packboards, the odd one on a horse. When the mail bag came off the train, the lodge-owner (whose several semi-official functions included that of being postmaster), would open it, and then call out the names of the addressees, who were all clustered around the stove in the waiting room. e.g. "Pete" or "Hank" or "Big John"; and when there was a letter for the timekeeper he would give an impromptu rendering of the "Last Post".

Mrs Weedon lived on a hill above McGillivray Falls, and one evening we paid her a visit. The trail switchbacked up alongside a steep set of cable car tracks (the "go-devil") - which explained how the lumber had been hoisted up for the frame house, and also how several unexpected items, such as a piano and sundry other very elegant pieces of furniture had found their way up there. Mrs Weedon explained with a twinkle that she was really a very cultured lady, who had been brought up to speak French, play the piano and observe the importance of etiquette. But - and here the twinkle turned into unrestrained laughter - if one is going to be a miner, one has to forego all these things. I remarked that she seemed to be doing well in both worlds - which indeed she was.

Lady prospectors were not unknown in British Columbia, but she was the first one I met. Many years later, in the late 1960's, Alec and I met her again, when she was living in a small cabin by the side of the highline trail where it crossed McGillivray Creek. I think by then she was in her nineties, and she had been burned out from her fine house down below. But she made a cup of tea for us, and still talked undaunted of the Golden Contact mine and her other holdings.

Camp life for me was now becoming routine. The novelty was wearing off, and the future seemed only too predictable. Ed, our instrumentman was a much older man than any of the assistants, who at one time were three in number, and there was very little assisting to do. So it was a very welcome letter that I received in March, from a Vancouver City survey firm, to whom I had earlier applied for a job. They offered me $120 per month as a chainman, which was a little less than I was getting at the time, however, I sensed definite prospects ahead. So, once again I gave in my notice, and duly climbed aboard the train to Vancouver.

STUDENT 1

"Circumstances rule man; man does not rule circumstances".

– Herodotus (c 485–c 425 BC)

Being employed by a firm of professional surveyors might be construed as putting ones foot on the bottom rung of the surveying ladder. I soon found out that unless I kept up to the mark in every way, I would be in danger of falling off even the lowest rung.

Looking back on that period of time, I think that the apprenticeship was a "rough ride". But I am certainly grateful to those partners in the firm for which I worked, for this so-called rough ride, which was simply due to the exacting standards which they demanded. It was standard practice, when training pupils, to "give them hell" all the time, and if such training convinced them that only the utmost precision was acceptable, then the treatment was successful. Moreover, such training as was received would - inevitably - be passed on to ones own pupils (for their own benefit and future prospects).

The firm was located in the old Williams Building, in downtown Vancouver. The building is no longer there, but the firm is still one of the larger surveying firms in British Columbia. I was duly issued with a plumb-bob, axe, box of nails, marking crayon and other assorted goodies which are tools of the trade for a survey assistant. I was of little use for the first few days, but picked up as best I could. I should say that two of the other assistants had been with the firm in such capacity for many years, and hardly a word was spoken while the work progressed; the surveyor and his helpers were so clued-in to one another's movements, that the job proceeded automatically. In a few weeks, however, the reasons for the various procedures became apparent, and I became a proficient assistant.

The preliminary examination of the B.C. Land Surveyors Corporation took place in April. All the papers were on mathematics, or related thereto. I took time off to attend the exam in Victoria, and passed successfully.

After the examination results were known, one of the partners handed me a printed form, which, at first glance, seemed to be a contract for voluntary entry into slavery. However, a careful second look at it confirmed that it was not as bad as it first appeared. These were my "articles", the purpose of which were to make an agreement between master and pupil, for passing on and learning professional skills. So, after duly filling in my name, and my boss's name at least one hundred times, as "parties of the first part" and "parties of the second part", I signed and watched my boss sign as well. I had started three years of apprenticeship. This turned out to be no different than simply working for wages, and learning as I went along.

Most of our work was in the city or suburbs. These jobs would be of one or two days duration, seldom more. I should add that there was, and still is, a great deal of "know-how" connected with city surveying, not the least of which has to do with working on streets crowded with pedestrians. I never actually saw a steel tape run over by a street-car, but it seemed that this was not due to lack of effort on the part of the street-car drivers.

I recall a few out-of-town surveys in that year also. One was in the Cariboo, and began with a 6 a.m. start by automobile to proceed up the old Fraser Canyon Highway. There are still traces of this road to be seen. Just north of Yale, it's single lane overhung the Fraser River on a wooden trestle alongside the C.P.R.tracks (this is where we met a bus coming the other way) - it crossed the canyon by the old Alexandra Bridge - and from there it weaved and switch-backed its way to Spencer's

Bridge, where it crossed again, following more or less the same route as the present highway. Driving this road took time. We stopped at the Alexandra Lodge for a meal; but it was not till 10 p.m. that we arrived at Clinton, where we spent the first night in the old hotel (since burned down). The following day we were picked up by a rancher, and driven by back roads, across the Fraser by the reaction ferry, towards the Big Bar country. There we duly set up our camp, and with two Indians as extra helpers prepared to lay out another 640 acres for the rancher. I liked this "cowboy country"; there was timber, swamp and open country by turns. There were even wild horses in the area. They had a disconcerting habit of licking the tops of our pickets with their tongues (for the salt), a habit which I wouldn't have objected to, but for the fact that I would then have to backtrack half-a-mile or more to drive them away, and reset the picket.

This work was, of course, quite different from survey work in Vancouver. But it was an education in itself, particularly in the art of communication between survey crew members. In those days there were no hand-held radios or cell-phones, therefore the instrument-man and picketman, when out of voice range, had to rely entirely on hand signals when extending the alignment. With a radio - as today - there is no problem. The instrument-man murmurs: "A little to your right - more - a little left - GOOD!", and the response is instantaneous from the picketman, who is listening in on his receiving set.

Our method required a lot more skill. "A little to the right" was signalled by waving a hat, or hand, to the right, and vice-versa to the left; while in between movements, the picketman would move his sighting card and plumb-bob as directed, at the same time stealing quick glances towards the instrument-man, to observe that all was OK. When the instrument-man was satisfied that the card was "on line", he would wave both hands frantically above his head, and the picketman would hurriedly mark the point on the wooden hub (previously driven into the ground). When the *second* shot was signalled (and there had to be two), the procedure would be the same, and one hoped that the point so obtained fell very close to the first. Because if it didn't, the picketman would have to signal a "wash-out" (both hands waved below the waist);

in acknowledgement of which the instrument-man would promptly throw his hat upon the ground, and jump on it. After which the whole procedure would be repeated.

Another out-of-town job was in Ucluelet, on the West Coast, and - again to show that transportation then took a lot longer than it does today- in order to get there one went to Nanaimo by the C.P.R .boat the afternoon before, and stayed at the Malaspina Hotel. Next morning early by Island Stage Lines to Alberni, transhipping baggage at each change, thence by the M.V.U-Chuck (an early one of the series) down the Alberni Canal, arriving about 8 p.m. in Ucluelet. Where, after putting our beds down, and eating a meal, Jim (who was one of my bosses) lost no time in arranging a game of bridge. It was not hard to arrange - there were but two employees at the cannery who played bridge, and both of them lived in eager hopes that two other players might happen to blow in. Jim himself was a keen and volatile bridge-player; in fact his keenness and volatility spilled out over most of his endeavors. I remember him with affection and admiration. His favorite expression when roused - "Holy old blueface!" - was guaranteed to lend color to any discussion, and nowhere to more effect than in the post-mortem after a bridge hand.

Office calculations for surveys were all done by logarithms. Although there were some mechanical calculators available, the use of them was discouraged by Jim, who actively doubted that (a) such a method could be quicker than logarithms and (b) that the right answer could result. We even had a race one day - he with a log book, myself with a hand-cranked calculator. However, although I won the race, he was still not convinced; until some years later, when the computer age began.

Nowadays, the searching of Land Title Office records is done mostly by "search" companies, who have desk space in the same building, and who can "fax" the data to their clients. The Land Title data is for the most part computerized and on microfilm. Land surveyors usually do their own searching, as they require more specialized information regarding parcels of land; much of this involves backtracking through old plans to ascertain the location of boundaries.

Searching in the old Land Registry Office in New Westminster was at first an intriguing experience. The lower floor was like a rabbit-warren, the corridors lined with old (and musty) files and document holders. One expected to meet Charles Dickens himself at any moment. All of this data is now (we trust) recorded on microfilm. But seeing the information today on a viewing screen is hardly comparable with making a search in 1947, when one fished out the original document, blew the dust off it, then made notes about it before replacing it. Searching today is *too* easy.

I spent about six weeks *copying* old plans here in preparation for a retracement of the old V.V. & E. Railway right-of-way. Copies of the plans (there were about three hundred) cost $1 apiece, but at my low wages it was obviously more economical for me to copy them. At any rate, by the time I had finished, I knew every nook and cranny of that old building.

After this session, it was a relief to be able to get out in the field once more, even if it was November. This is a rainy month in the Lower Mainland, and that year it was memorable because my other boss Clare, had devised a method whereby he could stand by the transit in the rain, and not get wet. It consisted of a length of half-inch pipe which could be driven into the ground alongside the transit, while, inserted and resting lightly in the top, was a large golfing umbrella with a straight handle. The purpose of this mechanism was, of course, to keep the transit dry (not the operator); but in fact it did neither.

Compared with the efficiency of the rest of our surveying methods, this was a dead loss. One gust of wind, and it was gone. I spent more time chasing the contraption down the street than I did in actual measurement. In Richmond it ended up in a dyke, where I nearly fell in whilst retrieving it; and on two occasions when set up on the centre-line of the tracks, it took off just as the Great Northern streamliner, with its revolving eye, came round the curve - it only came through twice a day, but it came *fast* - in which case it seemed more important to get the transit (and ourselves) out of the way first. Eventually, the umbrella was run over by a truck. Even the boss conceded that if it was going to rain in the future, he would just have to get wet.

Next spring, I was scheduled to be out-of-town all summer - an entirely agreeable arrangement, since, at that time, the Federal government had not yet decided that the cost of free room and board should be added to ones income. This would be the second season for surveys on the Columbia Cellulose millsite at Prince Rupert, which my employers were undertaking. The previous summer had been spent on preliminary work; this year there would be location surveys for pipe-line, dam and millsite, foreshore leases for booming grounds, tunnel alignment, and legal surveys for the whole development. Accordingly, my boss and myself climbed aboard the S.S.Catala, and had ourselves and our baggage transported to Prince Rupert.

We lived in the office building on the old Port Edward dock (since burned down). There was lots of space for living accommodation, and we had our meals at the (very) "greasy spoon" at Port Edward. By day we went out into the bush, rock and muskeg (for which Prince Rupert is famous), ran our lines and took our elevations, while all the time that other commodity for which Prince Rupert is famous, and against which no kind of rain-gear is proof, kept up its monotonous patter on our heads. We became expert at doing everything in the rain. But we never really got so well accustomed to it as the local people, who, it was said, grew webbed feet at an early age.

For relaxation we played cribbage, or read novels, or listened to the radio. Occasionally other men connected with the development, diamond drillers and timber cruisers, came to stay. One of the latter, a tall, muscular man with a mouthful of black teeth, who, it was acknowledged, went so fast in the bush that no assistant could keep up with him, is worthy of mention in this journal. He used to come in every two weeks, and when he got back to town, he would buy several bottles of rye, then head for the offices on our dock, where he was allotted a bedroom. I remember one Sunday, sitting in my room, through the open door of which I could see, framed in the half-open door of Howard's room, the head of his bed and a small table beside it, on which stood a tumbler full of what I naively imagined to be whiskey and ginger-ale. Every five minutes or so, an arm would reach out, grab the tumbler, then replace it half-empty, or empty. If it were empty, it was promptly refilled. As I later found out, it was neat whiskey, and the first bottle went down pretty quickly, after which the tempo slowed only slightly until all the liquor was gone. Then, the "warming-up" session over, Howard would depart for Prince Rupert, quite unaffected.

When Howard went to Prince Rupert, he did not spend his three or four days carousing from bar to bar; he went to Metlakatla, or Digby Island to the Reserve, where he had many friends and was well respected. Nowadays such an activity would hardly raise an eyebrow, but at that time there was still prejudice, and I even remember him being referred to as 'a bit of a squaw-man' - a most unacceptable expression today. Not that this bothered Howard. He was his own man, and one of only a few who, without a trace of discrimination, were prepared to associate with the local native people in the area; an attitude with which I for one agreed.

Towards the end of the summer, he had an accident in his gas-boat, when fuel in the bilges exploded. He was badly burned, and taken to Prince Rupert hospital. As a patient he was far from lonely. Whole Indian families stayed constantly by his bedside.

STUDENT 2.

"Man's best possession is a sympathetic wife".

– Euripides (485–406 BC).

Since coming to Canada, my experience with the opposite sex had been almost non-existent, due to the kind of life I had been leading. There were few women in the camps, and none at all in the surveying profession. Not that I was accustomed to spending much time in their company. I had lived for a long time in a man's world, attending all-male schools, then a six-and-a-half year stint in the armed forces. I must say that I felt more at home in the company of my own sex, who were basically of two types - those that one could get along with, and the other kind.

Women, however, were much different in social attitudes, and I had often found that behavior, which I had come to regard as normal amongst ones own sex, left a lot to be desired when in the company of the other one. But although I did not often go out of my way to look for such society, I was shortly to become well involved.

About this time, in 1948, I found out how not to buy a used car. I was now doing the work of an instrumentman for most of the time, so, with a view to convenience, I decided to get a vehicle which I could use both for myself, and - if necessary - for my work as well. My choice fell on an ex-army jeep, which I bought for the (then) large sum of $600. New cars at the time were still hard to get, and it was still possible to buy a used car, run it for a while, and then sell it for more than one had paid for it; although to be certain of success in the matter, it would be advisable to buy a car in reasonable condition.

After I had bought the jeep, I found that it had a cracked transfer case, which leaked a fair amount of oil. I was advised *not* to have it fixed, as this might cost more than the whole vehicle was worth. So I parked it on the street in a different place each night, so as to distribute as evenly as possible the pool of oil which was always left underneath when I drove it away. I used to "maintain" the vehicle myself. There was a garage on Alma Road which, for $5, allowed the use of their hydraulic jack and other tools while one did ones work; then charged for the lubricants used only. My bill for lubricants was high - the transfer case took six squirts of ultra heavy grease every few days. And since all of the grease didn't get successfully into the working parts, some of it found its way over me, my clothes, hair and fingernails, where it indicated a perpetual state of mourning. However, that jeep provided us with a lot of fun, and in the snow of that winter the four-wheel drive was very useful. I say "us" because I had taken lodgings near Alma Road, where there were seven or eight other male boarders, all of about my age; there was company all the time, and the jeep was in constant use.

The winter was cold. In Vancouver the temperature fell to minus two degrees Fahrenheit. At Fraserview (the second major veterans housing project - the first had been Renfrew Heights), my task was to measure all existing buildings, then plot them on a base map, My helper was an old hand named Gordon. We would work for one and a half hours, then head for a small cafe, where we downed a bowl of hot soup; then out to do another couple of hours work, before returning for more hot soup. We consumed a lot of soup this way. Minus 2 F is not very cold in some parts of Canada, but in damp Vancouver, with a wind blowing, ice forming on the foreshore at English Bay and steam arising from it, it was cold enough.

Sometime that winter, I met a Yorkshire lass, who had brown eyes and auburn hair; it was a stunning combination, and "involvement" started forthwith. We met through the auspices of "the family", Lee

Redman being our mutual cousin; and in no time at all, we found ourselves going out together. Kitty and her mother had a small apartment near Oak Street, close to St. Vincent's hospital, where she worked as a nurse. She had been in the Q.A's (Queen Alexandra's Imperial Military Nursing Service) during the war, and in the same theatre of war as myself, so we had a lot in common. The first time that my friend Bob and I visited her apartment for dinner, I brought three dozen daffodils from a corner grocery store. Naturally there were exclamations of delight at this tribute; had I been a real lady's man, I would then have let the matter rest. However, since I was simply myself, I then spoiled the effect by announcing our reason for buying them (i.e. they were so cheap that we couldn't resist). I have often wondered if my flair for making the wrong remark was the cause of my wife's approval.

In my youth I had always been an admirer of Rabbie Burns. I do not think anyone who has studied the philosophy inherent in his works could fail to admire him. However, I had never attended a Burn's Nicht dinner. So when Kitty and I were invited to one sponsored by the order of the Eastern Star, we were keen to accept. Two events took place that evening to make it memorable.

The first event occurred during the "piping-in" of the haggis, when one of the "great chieftains of the pudding race" unaccountably slipped off the waiter's platter, slithered across the floor, and exploded against the wall. The immortal bard himself couldn't have arranged for better entertainment.

The second event was also a happy one, though in a different context. I proposed, and was accepted.

Of course, the two events are only associated in my mind because they happened on the same evening. But it is interesting to look back and ponder the fact that, the night on which we plighted our troth was the night on which the haggis exploded.

Sometime in April, I was warned that I'd be going up to Prince Rupert again, for the summer. Preparations therefore had to be made for the sale of the jeep. These proved to be easier than anticipated. After explaining to the prospective buyer, who was a farmer from Merritt,

B.C., all about the cracked transfer-case, I took him for a ride. Half-way up the steep hill on 16th Avenue, just east of Alma Road, we had what can only be described as a contretemps. The back axle simply broke, with a loud crack.

I had never heard a rear-axle break before, yet, surprisingly enough, as soon as I heard it, I knew what it was. ("That", I said to myself, "is the back axle breaking"). Out loud I said:

"------------------!"

My client said nothing at all, but was surprisingly calm; and even more surprisingly still interested. I had the vehicle towed away to a garage, who quoted me $30 to have a new axle installed. It was the same garage where I had been doing my "maintainance", and I suspect they had been waiting for something like this to happen. (I make no comment on the $30 amount, which today seems completely ludicrous). Two days later, the buyer called on me, gave me a cheque for $625, and drove the vehicle away.

Early in May, Hamish and I climbed aboard another Union S.S. boat en route for Prince Rupert again. Hamish was my helper on this trip, and thirteen years later we would become partners.

PULP-MILL.

"Don't look back. Something may be gaining on you".

– Leroy (Satchel) Paige (c 1900–).

Prince Rupert in 1949 was all hustle. The pulp-mill which had for so long been expected, was now about to become a reality. Many of the residents found this hard to believe, their cynicism had been nurtured for twenty years.

The Company had by now hired a number of permanent staff, which included some of the engineers and surveyors who had worked on the project the previous year. They were installed in offices in Prince Rupert, and on the proposed mill- site at Port Edward. Our small party, however, was independent of the growing organization which would soon be supervising the construction in all its phases.

My first job was to rent a vehicle, which was done with the aid - I might even say connivance - of one of the permanent staff engineers, Roy. It was an old panel truck; the rent was reasonable; and it ran quite well while we did our work for a month; yet, looking back, I think I could have made a better choice. When I went to pick up my boss from the airport later, it chose that particular time to break down, locked itself in gear, and had to be towed away. Clare was often wont to be profane in a good-natured way, but this time he excelled himself in a most colorful fashion. We immediately made arrangements to rent another vehicle.

The work which we had been sent up to do took only a month. We checked the alignments of three proposed tunnels, and went on to do foreshore leases for booming grounds. One of the latter was on the south shore of the Skeena River. We drove up the road on the north bank with a small eight-foot dinghy on the back of the vehicle. The preliminary work consisted of triangulating to an island in midstream, then triangulating from the far shore back to the island. The Skeena must have been half-a-mile wide at this point, and in the spring it had considerable motion, carrying down logs, stumps and anything that floated, even whole trees. Yet my helper Hamish and I, undaunted - ignorance obviously being bliss, because we both thought we were doing a great job (which we were) - launched the little boat three-quarters of a mile upstream from the place on the opposite shore, where we intended to land, rowed across without mishap, and completed the work. When my boss came up later to inspect, and do more work on the site, he insisted on a bigger boat; in fact, he would settle for nothing less than a sixteen footer with a good inboard motor.

After this work was finished, my boss was going back to Vancouver, and so were we. However, construction having now started, there was a shortage of good instrumentmen at the site; the Company suggested to both my boss and myself that - maybe - I would like to transfer to the Company payroll, and stay. As I had been intending to get married in the fall, the idea was a good one - (a) I would get a raise in pay, and (b) I could save some of my earnings, as I would be living in the camp. The salary offered was a great deal more than I had been getting up till now, and it was obvious that I was welcome to stay, so I did.

The chief engineer's name was Klotz, an American, and his straw-boss was one MacAree. There was, as always, a certain amount of dissension amongst the engineering hierarchy. But in the lower echelons, all of last year's survey crew were there, some in different capacities - inspectors, expediters and others - and there was a good camaraderie amongst us. Most names I have forgotten, but there was an array of nicknames. Some of these were positively insulting, and only applied indirectly to the subject. But there were some fairly harmless ones such as:- "Lard-ass" and "Twitchell-puss" - and one man who had a hearing-aid was labeled "Old Wired-for-Sound".

Prior to my being employed by the Company, the first "pour" of concrete had taken place. Unfortunately the measurements had not been checked independently before doing so, and afterwards discrepancies were found of up to four inches; so the next day the jack-hammers were at work drilling much of the concrete out. The fact that the mill-construction crew was now short of an instrumentman who had been fired, was probably the reason that I had been hired. So it was natural that there should be heavy emphasis on precision of measurement at the site. We had here no less than three instrumentmen, and we even checked each other's work. There was talk of using invar tapes and other wild ideas; although the only important criterion was that the steel tapes which we used should match the standards used by the manufacturer of the structural steel. Comparisons were eventually made, one of our own tapes meeting the requirement as a standard, and it was noteworthy that one of the tapes we had been using differed by half-an-inch in three hundred feet.

Pipeline construction was also taking place from the dam on the Kloiya River down through three tunnels to the plant-site, and since I had done much of the survey work the previous year, Roy, the engineer asked me to look after grades and alignment. We had a little red truck, and crew of three, and spent our time driving from one project to another. When it rained - which was often - we would try to arrange to be underground.

That was not always possible, as, with the tunnels being driven on contract, and paid for by the foot, the miners worked a tight schedule. As soon as drilling and blasting had been done, the mucking machine went in, and when the muckers had finished, and before the next shift of miners went in to drill, that was when we were expected to set alignment and grade. We might have five minutes, and for the first three we would be peering through dust clouds. Eventually, as the dust cleared, we would be able (thanks to nineteenth century technology in the form of the carbide lamp) to distinguish a target or backsight, for long enough to make the necessary measurements. Carbide lamps which require just the right amount of water - no more, no less - could be temperamental, and unfortunately only one of our crew proved to be adept at handling them. If he wasn't present then the chances were that one of the lights would go out just at a crucial moment, stimulating many rich and colorful variations on the English language.

Another characteristic of miners was their measureless capacity for alcohol. They made on contract a great deal of money, but judging by the cases of whiskey that they consumed when off shift, I think they blew it as fast as they made it. But I never failed to admire their competence, (except when they argued that their version of "line" and "grade" was better than mine).

The tunnels, driven from both ends duly met as per regulations. I remember reading a standard text-book on underground surveying, which mentioned that "tunnels traditionally should meet on a dime". The same paragraph then went on to say that there had been times in the past when tunnels had not met at all. The writer quoted the case of a curved tunnel, where the instrumentman turned all his deflection angles to the right instead of to the left. He did not say what happened to that unfortunate instrumentman. . . . . I must say that the very idea of such a circumstance was enough to make me cringe - but for his sake I hope that he left unobtrusively in the middle of the night, and was at least three provinces distant by the time the error was discovered.

One of the men in camp was a Nova Scotia Irishman. That is to say, he came from Nova Scotia, and had an Irish name, but talked as though he came from County Cork. He was very happy-go-lucky. Towards the end of the summer, I had taken steps to rent a small apartment in Prince Rupert, as, after being married, I would continue to work on the construction while my wife and I lived in town. Apartments were hard to get, so when one became vacant I rented it, even though it was several weeks before we would need it. The apartment was in the south part of Prince Rupert, in a two-story building with a long corridor on the second floor, and flight of stairs down to the front door.

The owner, a pleasant man, spent a lot of time perched on a ladder painting the siding. It must have been a tedious job painting that big expanse; however, he broke the monotony by frequent referral to a bottle which he kept in a brown paper bag beside him; and by the afternoon was often in such an expansive mood that, on returning from work, I found it hard to get past him.

All my friends advised me that I should have a house-warming party, so we duly set a date for one Saturday night. It was a very good party. We drank a fair amount of beer. We brought in some food - the details are somewhat hazy - but I know that we weren't in the least rowdy. Until the time came to go home, that is. By then it was obvious that our Irish friend was well-looped. He was quite amiable - at first - and in fact all would have probably been well if he *hadn't* been so amiable. Because he insisted on walking backwards along the corridor, keeping up a running commentary: "Well, so long, fellows - boy! That sure was a swell party - see you tomorrow, eh? Boy! That sure was a swell. . . . ", and so on. The further along the corridor he got, the louder he had to say all this, so that I, who was standing at my apartment door would be sure to hear. So it should have been no surprise to us, when the door of the owner's apartment opened, and the owner appeared (in his nightshirt, and obviously propelled from the rear), and told him in no less a loud voice to "shut-up" as he was disturbing the tenants.

Now, no Irishman, even a small one, which our friend was, likes to be spoken to in that fashion, and he advanced on my landlord with a look of outrage, and both fists up. I hate to think what the outcome would have been, had not Harry, one of our crew members, grabbed him by the collar, yanked him to the stairwell, and given him a push. He bounced three times on the way down, and landed against the door out cold, with a crash that shook the whole building. Next day he professed ignorance of the whole affair, although everyone else remembered it with hilarity. Of course my landlord remembered it too, but the bottle I offered as a bond, to ensure that it wouldn't happen again, was accepted.

Sometime near the end of August, I took two weeks leave of absence, and headed for Vancouver. Kitty and I were married in Christchurch

Cathedral, a small wedding with our relatives and a few close friends present. The reception was at the Georgia Hotel. My friend Bob Greig had been slated to be best man, but a week before he was due to come to Vancouver, he was press-ganged into fighting a forest fire. It was quite a worrying time (apart from the usual worries that occur just before a wedding), and when he hadn't showed up by the day before the ceremony, we found a substitute best man; who did very well indeed. After the wedding, we left for a week at Yellowpoint Lodge on Vancouver Island, where we swam, fished, played tennis, rode horseback and also did a little relaxing. For me, this was the first holiday that I'd enjoyed since first coming to British Columbia. Bob showed up when we got back to town, with a set of dishes.

Soon after this, we again left for Prince Rupert. This time we flew by D.C.3. to Sandspit, on the Queen Charlotte Islands, thence by Canso amphibian to Seal Cove. I remember well the Canso, a WWII veteran , which landed on the water with a great swish, before climbing laboriously up the ramp to the terminal. On occasion (when it rained - which it did a lot) some of the local liquid sunshine would percolate through the fuselage onto an unsuspecting passenger, but I do not remember that we ever actually complained - getting wet was par for the course in Prince Rupert. This was now the standard route by air, until the airport was built on Digby Island.

I now joined the throng of workers who travelled from Prince Rupert to Port Edward and back each working day. A number of used buses had been bought by the Company for this purpose, and most of the time they ran smoothly; but breakdowns did occur, A mental picture persists (in black and white tones), of 3rd St. and McBride at 6 a.m., with the rain pouring down, and the street lamps reflecting off the black rubber clothing and sou-westers of workmen silently trudging to the bus terminal. Rain clothes were standard uniform. Kitty would usually leave for work at the same time as I did, and trudge through the rain, too, to her job at the Prince Rupert Hospital where her experience as a "theatre sister" before coming to Canada was well recognized.

New Year's Eve that year was celebrated at the Aero Club (a drafty hangar at Seal Cove), with the weather outside being minus two Fahrenheit. We men wore, if you can believe it, tuxedos, and the women were glamorous in long strapless dresses. I distinctly remember emerging from the dance to find the car lights just glimmering, and the battery dead flat.

The month of March marked the end of our sojourn in Prince Rupert, when Kitty and I travelled south, back once again to Vancouver. We can still look back with some nostalgia to our stay in what was always a very friendly city. But oh how it did rain!

BOUNDARY SURVEY.

"No one knows what he can do till he tries".

– Publilius Syrus (c 100 B.C.)

Sundry pairs of budding surveyors diligently measured angles and distances, and took observations on the sun (which conveniently appeared at the right time). No two flag poles in British Columbia were ever so painstakingly positioned as were these particular ones.

It was April 1950, and the field test for the final examination of the B.C. Land Surveyors Corporation was taking place in Beacon Hill Park, Victoria. The written papers on practical surveying, laws and regulations, astronomy and other necessary subjects, which occupied three days, had been completed the day before. So most students found it a relief to get out into the fresh air once more.

Various serious-looking members of the Board stood and watched us critically; and I do remember one particular student, whose hoarse bellow to his assistant (who was twelve feet distant from him), was loud enough to shatter glass at half-a-mile. Presumably this was for the benefit of the examiners, and if so, the strategy worked. He passed with flying colors.

Early in May that year, I learned that I had passed the exams, and my boss, who was President of the Board, duly swore me in; so that I became a fully-fledged land surveyor, licensed to perform legal surveys in British

Columbia. However, although I found the prestige easy to bear, putting myself to work proved harder. My parent firm had no large projects yet, and there was not too much other survey work going on in the Province that spring. Then I heard that the B.C./Alberta Boundary Commission required a qualified man for a large-scale summer job.

This work was to continue with the extension of the 120th meridian of West Longtitude. The last such field survey had been in 1924, when R.W.Cautley D.L.S. had run the line to a point some sixty miles north of Fort St. John; where, since there seemed at the time no practical reason for proceeding farther, it terminated. With the discovery, at the end of World War II of oil in North-East B.C. and North-West Alberta, it became important to both Provinces to know where their boundaries lay on the ground.

I was interviewed in Victoria by the Chief-of-Party, the late Bob Thistlethwaite, who was shortly afterwards to become Surveyor-General of Canada. Bob had the advantage of being licensed to practice as a surveyor in B.C., Alberta and the Dominion of Canada. My job would consist of maintaining alignment, controlled by precise observations on Polaris. In fact the regulations called for particularly precise observations - and a line that was within 3 seconds of astronomic North; expertese that I had to master in short order. I joined the crew at Fort St. John after flying up there, in May 1950.

My wife was now working full time at St. Vincent's Hospital, again in the operating room. We discussed what this enforced and prolonged absence would mean - no letters or telephone calls for the whole summer. The crew of which I was a member would simply start at the end of the old boundary line, and hack our way northwards through the (literally) uncharted bush, for five months; returning in October, ragged, unshaven, dirty and smelling of pine needles. We decided that beggars could not be choosers. I could not turn down available work; after all, what was a surveyor for, if not to go out in the bush. (A few years later we all became "sidewalk surveyors"; but at the time of which I write, most of ones work was in remote places).

The crew was concentrating by the Beaton River, just outside of Fort St. John. There were twenty-one of us, including four packers, a cook and flunkey. There were thirty-two horses - twenty-eight of them pack horses. The rest of the crew consisted of a levelman and his assistant, three chainmen, instrumentman (myself), picketman, four or five axemen, and a moundsman, whose job it was to dig the pits and mounds which were to be set beside every survey monument, in order to comply with the Dominion Land Act.

The first few days in camp were spent in constructing rafts to get twenty-eight pack loads of gear across the river. When that had been accomplished, we started a long trek northwards in order to find the end of the old 1924 line. We followed an old trail - the Donnas Trail - little used by then. Our crew were mostly from Fort St. John. Angus Beaton, the boss packer was a very stalwart man; the Beaton River had been named after his father, who had been the Hudson's Bay Company factor in Fort St. John many years before. He was an excellent man in the bush. His brother, however, also a packer was less excellent. The levelman, John, and head chainman were young university students. All the rest of the crew had been locally recruited from the Fort St. John area, including the moundsman, who was an ex-miner, originally from Glasgow. This last member of the crew had an endless fund of lurid jokes, and kept us in stitches for at least the first part of the trek. By that time, we had all contributed our share, and the jokes became less funny, till after a while we were able - in a group - to supply the punch-line simultaneously with the raconteur.

It took us about six days to find the "old line". After following the Donnas Trail for three days, we met up with an Indian trapper. We added him to the crew as a guide, which was fortunate. The very next day the trail which we were following was completely wiped out by an extensive burn, which had taken place some ten years before.

So our Indian - who himself did not know the trail (although he had been told about it by his father) - took over as guide. For two days we followed him, sweating through the dry, blackened forest, cutting out logs for the horses; and by the end of the second day, there were definite murmurs of discontent from the rest of the crew. ("Where the hell are we getting to - this guy doesn't know where he is going. . . . " etc). But on the third day we were through the burn, and there in front of us was the old pack-trail, cleared through and blazed. To me - and the others - it was an eye opener; all of us were experienced in the bush, but our bushcraft was mere fumbling when compared to that of the Indian.

Yet for him - the Indian - the trail was all-important. He would not venture into the bush at random. Next day, figuring that we had travelled far enough northwards, we observed for latitude, and then, deciding that we should now head due East in order to find the "old line", started off through the bush in that direction. But the Indian would not come; we left him at the trail, and he returned to his trapline.

Having found the "old line", we commenced our northward survey, crossing the Chinchaga River early in the proceedings, and later on the Fontas, co-existing with insects, moving camp every few days. This was virgin country; there were no trails, no old stumps or signs of any former human activity; there was little game, and the few beaver which we did see had obviously never encountered a human being before.

Regulations decreed that the line must be "opened up" to a width of twenty feet at the top, which meant that we had to clear a swath about fifty feet wide at ground level. We had no power saws; the axe was king, both in camp and on the line. The boss had air-photos showing the route, and it was mostly flat ground. In the hot, summer sun there was a great deal of refraction, and the mosquitos were infinite. The boss had given us a talk on water conservation, and we had one large water-bag for the cutting crew; however, Scotty was a law unto himself. All along the line were to be found holes in the muskeg, one or two feet down to the perma-frost, which in half-an-hour would fill up with water. These were Scotty's water supply, and the wigglers contained in them were simply assurance (according to Scotty) that the water was pure. (The wigglers were actually mosquito larvae, and we soon learned to ignore them).

Caches of food had been established the year before by air, in the vicinity of the proposed route, and every three weeks or so the boss packer would lead a string of horses off to get grub. There were a number of interesting experiences that summer. Forest fires swept across the line behind us, and one went across in front of us. On two occasions we dug holes to bury our instruments, in case it became necessary to make a quick get-away. We had one camp where the water was bad, and half the crew became sick. And we ran into a peculiar type of terrain (known to surveyors as "loon-shit"), - which, when stepped on, breaks out in sympathetic surface-quakes three hundred feet or so distant.

We would work ahead until we were about three miles from camp. Then next day, we would pack up our personal gear and tent before leaving for work on the line. The packers would pack up the camp behind us, and move past us, to establish another camp about three miles ahead of where we were working, to which we would proceed at quitting time.

One such move was almost disastrous. The boss packer was away at a cache, and it fell to his brother to pack up the camp after we had left for work. After loading the horses, he proceeded up the line, past us as we worked, and onwards . . . . Bear in mind that this was flat country, interspersed with muskegs; but the sun always shone, and it should have been apparent to anyone whose head was screwed on the right way round, which way was north. However, what our friend did was to deviate to the east in order to avoid a muskeg, and then keep going east until quitting time, when he made camp. Fortunately for us, he lost a couple of horses in the process. We, the crew, after work proceeded north to the creek where we expected to find the camp; and found nothing. We built a big fire, but it was a cold and hungry night. At dawn we backtracked, and followed the horses' trail. Without the delay due to those lost horses, there could have been a disaster, because on foot we could never have caught them up.

Later during the survey, we reached the south fork of the Hay River, where we set our last monument. We had run about 67 miles of boundary, although individually we probably walked upwards of five hundred miles each.

It was now getting to be fall weather, and we wended our way along the river bank on a *real* trail for two days, until we came to a lake. This was where Bob planned that we should be picked up by plane, after sending the horses back to Fort Nelson overland. On the way to the lake we lost one horse, the only one on the entire trip. He had been skin and bone for some little while, and had merely been following behind the pack train, but finally he fell down in a swamp, and could not be got up, to the disappointment of the boss packer who had hoped to bring him out alive. So Angus shot him, and the swamp became his last resting place, and even Scotty, who had been mumbling about horse-steaks for dinner, did not feel like violating it, and thus incurring Angus's displeasure.

At the lake we had a problem. Due to low-lying smoke from a multitude of forest fires, the plane could not locate us. Moreover, we were running out of food. I think we spent a week at that lake. Every time we heard a plane, John would put on his good pair of pants (which he had been saving for the occasion), and throw his old ones into the bush. Then when the plane flew away frustrated, he would retrieve the old ones and put them on again. While we were waiting, we played "baseball" with a ball made from the heel of a rubber boot; we also competed at making frying-pan bannock (flour being the only foodstuff not in short supply). We had two fishing lines, so we built a raft and caught jack-fish out of the lake. We also had three guns in the group, so the owners of them went out each day, in an ever increasing radius from the camp, and brought back prairie chickens or grouse. Our diet, which would otherwise have been simply beans and bannock, was thus varied. We had long since run out of bacon.

Eventually Bob was able to talk the pilot down to the lake, and John, Scotty and I climbed aboard the Norseman, which, after one false start (the pilot taxied over a mound of submerged boulders, ending up high and dry, but luckily with undamaged floats) - took off and landed us at Charlie Lake. Here, my first task was to order some food to be taken back for the rest of the crew. My own clothing was somewhat incongruous, consisting of "long-johns", partially covered by a shirt full of holes, ragged pants and I had a four month old beard. I am sure we smelled quite rank. Yet the girl behind the counter at the Charlie Lake store never batted an eyelid when we walked in; she looked us straight in the eye. I concluded that she must have been well used to the sight of such weird characters emerging from the bush.

Scotty had been telling us all what a bender he was going to have, as soon as he got back to Fort St. John. And sure enough, when John and I found the only hotel room available, there he was ensconced in the next room. All three drawers of his dressing table were filled with liquor, and Scotty - who by now had acquired many bosom friends - was in an expansive mood. I remember one of his friends reciting "The ballad of Eskimo Nell", while the outsize Indian woman who was sitting on his knee, and who, I am sure, understood not a word of that poetic masterpiece, hooted with laughter at the end of each line. Yet in spite of all this we managed to get to bed at a reasonable hour; Scotty, who had been drinking Drambuie out of the bottle ("My national drink, you know!"), passed out cold at about ten p.m., and the party petered out.

PRIVATE PRACTICE.

"Remove not the landmark on the boundary of the fields".

– The Wisdom of Amenemope (10th cent BC)

Vancouver was a new world - where people sat in chairs and lived in houses; where one could walk on the street side by side, instead of in single file, swatting mosquitos the while; and where there was other food besides bannock and beans.

But best of all, to enjoy Vancouver meant resuming married life again, and the company of my wife, who was about to show how adaptable she also could be. While I was away, she had been living with a girl-friend in our basement suite on Oak Street, working all the time at St. Vincent's Hospital, and by the time I returned she had her driving licence and a small Austin A40, in which she met me at the airport. The car lasted us well for several years, in spite of the fender ripping that took place while I was demonstrating to Kitty how to park it. The inference was plain - it was time I went back to the bush again. And in due course I did, but in a closer location, where I could get home for at least one day a week.

Whereas until now the tenor of my life had consisted of one haphazard incident after another, from now on events seemed to follow in a more well-ordered and logical sequence. This was undoubtedly due to the fact that I now had a home, and a wife. Most decisions would now be made more deliberately and with (so we thought) an eye to the future.

In the spring of 1951, I made a trip to the North Shore, and rented desk space, plus half a room for storage, in the back of a real-estate office on Lonsdale Avenue. My shingle went up in the window. Shortly after this, Kitty decided to quit nursing and become my "office staff". The switch was of great benefit. She kept the books, liaised with the Land Registry Office, answered the telephone, and did a hundred and one other necessary jobs. Actually my telephone seldom rang, but I was kept very busy through referrals from other surveyors. Of course, I made the rounds of government offices, Municipal engineers, real-estate firms, lawyers and others - a procedure that must be familiar to all surveyors starting out in private practice. British Columbia had not yet embarked on its post-war boom; when it did, there would be too much work rather than too little.

I spent two months in the North Shore mountains, defining the Nelson Creek and Cypress Creek watersheds, I also worked on the Walleach Transmission Line in Surrey, had a spell at Zeballos (where my wife paid me a visit aboard the S.S.McQuinna), the Skookumchuck Narrows, Port McNeill, Prince Rupert (again by Union S.S.), and did hardly any work in North Vancouver, where my office was located.

Twice I went to Alert Bay, the first time in an old Stranraer flying boat, courtesy Queen Charlotte Airlines, in which it seems I was lucky not to have used the toilet. According to Spillsbury's recent book, "The Accidental Airline"), Harbour Publishing 1988*, this aircraft was known as "the whistling shit-house" due to certain peculiarities in the toilet system.

The diversity of jobs was partly intentional (I *liked* being out-of-town), and partly because in-town work was scarce, all of it being undertaken by other well-established firms. One could hardly blame any such firm for preferring the sidewalks of Granville Street, to the fifteen-feet-long salal and rock bluffs of the B.C. Coast. For myself I was quite happy to work in any location, provided it was in British Columbia.

One small job I remember from that time consisted of flying to Elk Lake in Northern Vancouver Island, to make a contour survey of the top of a mountain for a mining company. The trail from the lake up which Jack, my helper, and I backpacked, led through an abandoned mining camp, with the bush growing through floor-boards and roofs laying askew - a sight that never fails to arouse in me a feeling of regret for all the hard endeavor that goes into providing living conditions in remote places. The trail ended up on top of a hill at about elev. 6000 ft. This was the prospect, and as several tents stood empty, Jack and I stowed our gear in one of them, then went in search of the cookhouse.

"Do you like "stinky" cheese?" - was our greeting from a fierce-looking character with a two month old growth of beard.

Now it happens that I do like "stinky" cheese, and replied in the affirmative, whereupon I was handed a large lump of Cheshire Blue and two pilot biscuits; but something about the way the question was asked, and the glint in the questioner's eye, told us that if we said "no" we'd be in trouble. Fortunately, Jack was also partial to "stinky cheese", so we forthwith lunched, while our host, who was the mining engineer in charge of the camp directed the dialogue.

As a topic of conversation "stinky cheese" has certain limitations, but we gave it our best, and for the first meal at any rate, we found it was a subject that we could all agree on. At other times we were treated to a second topic: -"I'd quit the son-of-a-bitching company tomorrow, but if it's the last thing I do, I want to see them make a mine . . . . . ." after which would often follow a monologue regarding ounces per ton of copper, confirmed or otherwise - most of which was over our heads, and needed no reply.

The camp consisted of the engineer himself, a diamond-drill crew, and several back-packers, who went up and down the trail all day. There was no lack of variety as regards food. Jack and I made several good meals ourselves. But the engineer stuck steadfastly to his chosen diet - "stinky" cheese; I can honestly say that I never saw him eat anything else, and he certainly had a craving for it, as the packers kept bringing up more. But he ate no meals as such, simply ate the cheese and smoked innumerable cigarettes.

The purpose of this anecdote is to demonstrate the real meaning of the expression "to be bushed". Compare the present-day application of the phrase, as used by an executive coming home after a hard day at the office, setting aside his brief-case, sipping his first martini, and saying "I'm bushed!". Fatigued he may be, but bushed he is *not*. But my engineer friend *was* bushed; he had been up there on that mountain-top too long.

Two months later, I met him walking along Howe Street; a most charming and almost different person. We chatted, and he gave me news of the company plans. And not a word about "stinky cheese"!

PIEBITER CREEK.

"Let each man exercise the art he knows"

– Aristophanes (c. 450–385 BC).

When I first met Mrs. Noel, she was acknowledged as the grand old lady of the Bridge River valley, a position she earned through having prospected and hunted throughout the area since the year 1900, when she first came to the valley with her husband Arthur Noel. She would certainly not have appreciated use of the adjective "old".

She was the daughter of French-Canadian parents in Lillooet, who sent her back to a Quebec convent for her schooling and she never lost the slight accent she had by then acquired. She and her husband were the original stakers of the Lorne Mine, later to become part of Bralorne, which at one time was the second largest producing gold-mine in the Commonwealth (including South Africa). Late in life she was awarded the Canada Medal, a well-deserved honor.

There is no doubt that she benefited substantially when Bralorne was financed (she would drive a pretty hard bargain, and I am sure she got a fair-sized "piece" of the action). Yet she chose to continue living the hard life. She told my wife that she had had five miscarriages, due to the strenuous life she led. In summer, she dressed in an old pair of breeches, mackinaw jacket, and an old battered hat that had seen better days, and roved around her mineral claims (at this time the Chalco Group at Piebiter Creek), collecting samples and prospecting. In the winter, she retired to the West End of Vancouver, where she played bridge for relaxation, and wore the same clothes that most old ladies of her vintage would wear - nylons, a dress, a hat transfixed with hatpins and carried a handbag. However, when she walked down the street, it was obvious that this was no ordinary little old lady; she strode out like a highland gillie. She told us a story about herself which proved the point. Some years before (she said) - she had been referred to a doctor at the Mayo clinic. Having been examined by this doctor, she was then passed on to another, who then referred her to yet another doctor; at which time she asked the last doctor what was afoot. One examination should have been quite enough, in her opinion. The doctor admitted that the second and third times had been unnecessary, but none of the doctors had seen such physical development in a woman before, and they wanted their colleagues to see it also. There could have been a lot of truth in this story.

One day in July 1951 we set out, a crew of three, to survey the first four mineral claims of the Chalco Group at Piebiter Creek. An intriguing name such as this begs for an explanation. This can be found in G.P.V. and Helen B. Akrigg's book - "1001 B.C. Place Names" (Discovery Press 1973)* which tells us that the creek was called after one "Piebiter Smith", an early prospector with protruding teeth and a fondness for pies.

I was fortunate in having Jack as my main assistant, a man whose hobby was mountaineering. Precipitous terrain, far from slowing him down, actually stimulated him; as did the prospect of seeing new vistas, as he was also a very competent amateur photographer. I have to admit that, in 1983, when revisiting Piebiter Creek, I found it hard to believe that anyone would willingly run lines in such steep, rough country; yet, in 1951 we did, and I do not remember that we found the work unduly arduous.

To drive to the Bridge River valley in 1952, it was necessary to ship one's car by rail from Lillooet to Shalath, to which end the P.G.E. Railway Company provided a flat-car service twice per week. So we drove to Lillooet, put our car on the flat-car to Shalath, then drove over Mission mountain and on to Bralorne, where Mrs. Noel awaited us. This took a day. Next day we drove through Pioneer, up to Piebiter Creek, a primitive road, but drive-able by an Austin A40. From the end of the road it was a two mile back-pack to the claims, and we got there with all our gear by the end of the second day. "There" was a group of log cabins and tent platforms. We stayed in tents, Mrs. Noel was in the main cabin, where she cooked our meals. I got the impression during that two week sojourn, that meat - to Mrs. Noel - meant steak. No other meat was considered edible. Not that we complained; we had never eaten so well in our lives. When supplies became short, she took her "Trapper Nelson" packboard, walked out to Bralorne and came back with more "meat".

The claims were at an elevation of 5500 feet plus - that is to say, all the way up to about 6500 feet, and they provided my first introduction to "slide-alder". This is descriptive of a small alder tree, ten to fifteen feet high, which grows at high elevations on steep slopes. The stem, up to four inches in diameter, grows horizontally outwards for about two feet, then straightens up and grows vertically. It is usually found on old rock slides, and sometimes stretches for half-a-mile. It makes for tough going.

But the problems of contending with slide-alder were negligible compared with sorting out the mineral claims. In those days, by definition, a mineral claim could not include land which was already lawfully held for mining purposes. The precept still holds good obviously. Now this perfectly logical and innocuous statement is fraught with hidden complexity. For instance, where, at the time of staking a claim, part of the land is already held by a prior claim in good standing, such ground must be excluded from the claim under consideration; even if the prior claim were to be abandoned a month afterwards, the claim under consideration would never be entitled to such ground, which (when later abandoned) becomes open ground, and must be re-staked. Although all surveyors are well aware of this ramification in the otherwise straightforward definition of a mineral claim, at the time of which I am writing there were even some Mining Recorders who were not wise to it, and certainly a great number of free miners.

Mrs. Noel, when staking the Chalco Group of mineral claims, started out in an area where there were no conflicting other claims. It was all "open" ground. She had an assistant, one W.E.Rutledge, whom she employed to help her stake. She could stake eight claims only in her own name, but could acquire others by purchase, so she asked her assistant to stake eight more, which she could then obtain by bill-of-sale for a nominal sum - $1. This was standard practice amongst prospectors, and was a recognized method of acquiring mineral claims. The assistant would agree at the start of the operation to hand over the claims which he had staked, although only bound to do so by a gentleman's agreement. Well - Mr. Rutledge turned out to be no gentleman; because after Mrs. Noel had staked her claims, and he had staked his (they were merged with each other, and considerably overlapped), and the claims had been recorded, Mr Rutledge decided that he would rather keep his claims than turn them over to her.

In Mrs. Noel's interesting career, several persons had tried unsuccessfully to get the better of her, and Rutledge must have been naive indeed to suppose that he could get away with this deal. She could do nothing legally, of course - the claims he had staked were officially his. But next year she staked more mineral claims around his holdings on the outside. Then she waited him out; she would not buy him out, as he had hoped. She simply held on, keeping her claims in good standing. Five years later Rutledge passed away, and his claims lapsed. So Mrs. Noel re-staked over his lapsed claims, and again over her own (some of which were in good standing, others never having been so). Then she decided that maybe she had better have a survey, to find out just what ground she really had covered; this was where I came into the picture. There were at least twenty four claims to sort out.

I could not ever have found a better introduction into the practical workings of the Mineral Act. It was the worst "dog's breakfast" that one could possibly foresee. Adding to the confusion was the fact that, since it was an area of many snowslides, a good many of the staking posts had disappeared, and I had to obtain affidavits to establish the positions of such missing posts. Fortunately, Mrs. Noel had a good memory. After more than a week, we were able to define legally and properly the ground to which she was entitled. Predictably, we found a large "hole" of open ground in the middle, and staked it on her behalf as a fraction.

The Sub-Mining-Recorder in Goldbridge at the time was Will Haylmore, also a well-known personality in the history of the Bridge River. At the time of recording this fraction, he had long hair down to his shoulders ,like Buffalo Bill; and we had a short discussion on fractional mineral claims - which might have gone on longer, as the whiskey bottle on the table was still half full. But regretfully we were on the way to Shalath to catch the "flat-car" to Lillooet. There is a good picture of Haylmore in the publication "Bridge River Gold" by Emma de Hullu, published by the Bridge River Valley Centennial Committee 1967*. It shows part of the circular rock wall outside his cabin, with one of several wooden models of Lewis guns, which were painted white and set in the wall facing outwards. His grave is still on the property, and is being tended, but the rest of the land, which was held as a placer mine, reverted to the Crown on his death, and the cabin with its pictures and regimental photographs of the British army of bygone days, was later pulled down.

There is another excellent photograph in the same publication. It is of Mrs. D.C.Noel as a young girl with a trophy - one of the many grizzly bears that fell to her gun.

BRIDGE RIVER AGAIN.

"It is good to live and learn".

– Miguel de Cervantes (1547–1616).

The following year, 1952, saw us back again in Mrs. Noel's country.

By now I had found that, in spite of the compensations of working in the city where I could live at home, there were drawbacks. Every client wanted his particular job done by tomorrow at the latest. This necessitated working on evenings and week-ends in order to catch up on the requisite office-work. Whereas, out-of-town conditions were usually less harried; jobs were of larger magnitude, and office work could then be done while the crew did work that did not require supervision. In 1952, I was away all summer by intent.

Now that her Chalco Group claims had been sorted out, Mrs. Noel had decided to stake and survey some more. This time we had a different arrangement. My wife, who otherwise would have sat in the office in North Vancouver, twiddling her metaphorical thumbs, thought that she would like to be part of the crew. This was an excellent idea, as we needed a cook. Later on in the summer, I was to do two jobs for the Ministry of Energy, Mines and Resources (Canada), and for that work we needed a book-keeper. My wife was able to fulfil both functions, and she also became the first female crew member to be employed by that Federal Government department on a legal survey.

We arrived at Piebiter Creek with all our baggage, and set up our own encampment about quarter-of-a-mile from Mrs. Noel's cabin, beside the creek. This time there were no problems of title to worry about, our claims were simply projected up the valley, and we spent the next two weeks working steadily at them. My wife had a cook stove set up in the open air (covered with sheet metal strips, after we were rained on). We had two tents, and a good camp. Bill, one of the crew, was a young student; the other, Rolfe, was a Norwegian, who made up for his lack of English by his knowledge of the bush. He was an excellent axeman.

Close to the camp there was a big rockslide. Mrs. Noel had on occasion remarked that this was where she wanted to put in a tunnel. Although I thought I knew what miners could accomplish, I must say that the prospect of tunnelling through a rockslide without machinery seemed to me at the least impracticable. But this lady knew otherwise. She went in to Bralorne one day, and came back with Oscar.

Oscar was a Swede, and I believe he was approaching eighty years of age. He was quite shaky, and when he walked, his knees had a way of knocking against each other. He slept in a tent in Mrs. Noel's camp; and while we were in the hills surveying, Oscar started work. He used only hand-tools - crowbars, hand-drills and dynamite, and he cut and squared his own timbers for the adit.

Napoleon once said that while a difficult problem can be solved right away, the impossible may take a little longer - or words to that effect. In three days Oscar, working entirely on his own, had a timbered portal ten feet long through the slide, and was drilling into the rock-face. We, non-miners that we were, stood and marvelled at it. If Oscar could do this at seventy-nine years old, what sort of a man must he have been in his prime.

But next day, he was gone; which Mrs. Noel had expected. She gave him another two days, then went to Bralorne and brought him back, completely incapable and incoherent, and when we had helped her put him to bed, she left one-third of a bottle of rum under his pillow - a very understanding lady.

Our work went well, and the only other incident that I remember of note, apart from all of us being beaten by Mrs. Noel at crib, was arriving on the line one morning to find that a porcupine had all but eaten one of the wooden transit legs. We improvised.

After Piebiter Creek, we proceeded to Little Gun Lake, after having met with the Bralorne Mine manager, Don Matheson. Our next job was to survey some lots fronting on the lake, for the company. We stayed in one of the cottages close to the Little Gun Lake Lodge. After the heat and mosquitos of Piebiter, this place was a Paradise. We swam every day, and fished enthusiastically if inexpertly. Mrs. Noel had advised us that Don Matheson was a very nice man, and that "he would do anything for me". While talking to Don, the conversation inevitably got around to her.

"Well, of course I'd do anything for her", he said, "I find that it *pays*. If she asked me to stand on my head here in the office, I'd do it. The fact is that if she wants something, you might as well realize at the start that she will not rest until she gets it; so in order to avoid a lot of hassle, I make sure that she gets it right away!"

Which statement does a lot to sum up Mrs. Noel's character. She was a lady of great charm, and strong will power. Frank R. Joubin, possibly Canada's most famous geologist, was a personal friend, and wrote a detailed appreciation of her entitled "Bridge River Pioneer"*, shortly after she was awarded the Canada Medal in 1958. She died some years later aged eighty. She left a house in Bralorne that was quite unique. It had a stone fireplace containing pieces of ore from all of the currently producing mines in B.C., and hanging on the walls were skins of several of the grizzly bears which she had shot. Unfortunately, the house was not taken care of, and it was later vandalized.

After finishing work at Piebiter Creek in 1952, I did not return to the area for four years. But the lure of the bridge River, its history, its mountains and lakes had already prevailed on me, and In 1956 I purchased a lot on Tyaughton Lake. What transpired was a long-lasting association - almost a love affair - with that unique part of B.C; leading inevitably to the building of a house and retirement; more of this in a later chapter.

INDIAN VILLAGES 1952.

"If you only go once round the room, you are wiser than he who sits still".

– Egyptian Proverb.

Kitkatla is a small island lying about sixty kilometers south-west of Prince Rupert. It was the home of the Kitkatla Band of First Nations people. Here, on behalf of the Indian Affairs Department of the Federal Government we were to do a short preliminary survey; for the purpose of future planning; after which we would proceed to Kitimat Indian Village, at the end of Douglas Channel, to undertake a larger project.

Kitty had now taken on an extra job, that of book-keeper. I had been advised by an older and wiser man, that if one were to take on work from the Federal government, one should first hire a book-keeper, then put him/her on the payroll. So - as I was informed later - Kitty unwittingly had the honour of becoming the first female employee to be hired by a crew doing work for the Federal Dept of Mines & Resources.

At that time the government were very picky. Copies of telegrams, subject matter of telephone calls, all had to be recorded; and when I submitted a grocery bill with listed amounts only, I was told that, in future, receipts would have to be itemized in full, for example:- one pot of jam @ $1.25: two tins of sardines @ $1.50 and so on . . . . . . Yes, we needed a book-keeper.

The crew - Rolfe, Bill and Kitty flew up to Prince Rupert direct via the Canso Flying Boat (a service recently instituted by C.P.A., but discontinued soon after). I went on a day ahead myself, to arrange accommodation, plane charter, and liaison with Indian Affairs and Land Registry Offices in Prince Rupert. By the time the crew arrived, the hotel rooms were reserved, plane chartered, and we spent the rest of the day buying food. A well-planned operation, you might say.

However, the best laid plans of mice and men - as Rabbie Burns so well put it - "gang aft aglee". A very descriptive phrase - trust Auld Rabbie to think of it. For the next few days, things could not have gone more "aglee" if Rabbie himself had been there to coin the expression.

Three times we checked out of the hotel, and waited - our gear all packed - on the wharf, but no float plane materialized. In between times, I paid visits to the Queen Charlotte Airlines office, where the manager's expression became more and more embarrassed each time he saw me (- I even felt embarrassed at making him feel embarrassed, as it was obviously not his fault that we were grounded). Prince Rupert weather is noted for its contrariness, and when it was clear for taking-off in Prince Rupert harbor, then there were four-foot waves at Kitkatla; and when landing at Kitkatla was safe, then Prince Rupert was fogged in.

So, after two days of waiting, I approached the Indian Affairs Department. They had a perfectly good (so we thought) boat, which they very kindly put at our disposal, also its pilot. An hour later we were tossing our baggage light-heartedly into this 30 ft. diesel launch - a very classy looking craft - fully expecting to be at our destination that evening. But we were again disappointed. Something was not quite right with the engines, and we had to put back again to harbor. My wife consigned the current evening meal to the deep - (two days of thawing out in the sun had made the T-bone steaks quite unappetizing) - and we prepared to unload and return to the hotel. However, the pilot assured us that we could stay the night on board, and the engines would be fixed by morning. Again full of hope, we bedded down; it was a comfortable boat with lots of room for all. In the morning we started out again. This time we actually got to the entrance to Prince Rupert harbor, where the engines quit for good, and we bobbed around - a considerable hazard to navigation - until the R.C.M.P. boat took us in tow back to the wharf. And there, back at "square one", we resignedly tossed our baggage back on the pier.

At this point, my wife reminded me that one of the doctors in Prince Rupert had a boat. It was this bit of knowledge - and the fact that the good doctor had a high opinion of her when she had worked at the Prince Rupert hospital two years before - that made possible our journey to Kitkatla. Within the hour, we had a good fast boat, and pilot (recommended by Dr. Green, our benefactor), and we were on our way. We arrived some two hours later, quite dazed by this change in fortune.

Had I been a regular employee of the Federal government, I probably would have thought little of this delay in travelling the forty odd sea miles from Prince Rupert to Kitkatla. But having been "brought up" - as it were - in private practice, where a survey has to be a paying proposition, I was horrified at having to take four whole days to do it. I still wonder how so many events could simultaneously conspire against us.

Most of the men in the village of Kitkatla were away, logging or fishing, as was the Chief; however, day to day affairs were in the capable hands of Mrs. Vickers, the Chief's wife. There was a main street of "board-walk". Water was supplied from a dammed-up creek, which was in actuality a swamp, via a wood-stave pipe, and taps which were strung out along the village street at 100-foot intervals. Had the taps ever been turned off, then some water might have been available, but the Natives were inclined to leave them open, consequently the dam was often dry. I remember seeing such taps in other Native villages, all of which had the same problem, and wondering why a government department, which had gone to the expense of providing a domestic water-supply in this fashion, could be so remiss as not to put in automatic shut-offs.

A surprise was in store for us when we arrived at the teacherage, which had been officially set aside for our accommodation. There were two teachers in residence, having an unofficial holiday, and there just wasn't enough room for all of us. For two days we shared the accommodation, in a state of polite hostility, with our two crewmen sleeping on the floor, till at last the teachers gave up, and managed to get themselves transported out on a local fish-boat.

Kitty, meanwhile, spent two afternoons exploring the headlands, and by way of helping Mrs Vickers, arranged to take two of her little boys on a "picnic"; one of them was undoubtedly Henry, who would soon be making his name as the most talented artist to depict the pageant of British Columbia.

When it came to be our turn to be transported out, providence was not so kind. We were not on any usual transportation route, moreover we wanted to go to Kitimat, rather than back to Prince Rupert. So we went through the same procedure as before of attempting to charter a plane. But this time we waited only two days, before cancelling in favor of other methods. I radio-phoned to a friend in Prince Rupert - Crawford M., who had been one of our survey crew at Port Edward three years before - and was therefore a bosom buddy. (He was now - appropriately - running a travel agency). When I begged him to "get us out of here" by whatever means he could, he chartered a tug for us.

Next morning at 6 a.m., we heard the boat's whistle at the wharf, and we lost no time getting aboard. It was a very pleasant day's voyage. Our skipper ran the boat single-handed in a most efficient way. We took turns at the wheel, there was coffee on the stove all the time, bacon and eggs whenever required. Outside it was a beautiful day, with a calm sea, and the unsurpassed scenery of blue water and green rocky islands. The British Columbia coast had donned its pleasantest garb.

By nightfall we were in the Douglas Channel, and about 8 p.m. we tied up at the Kitimat Indian village wharf. It was a very low tide - 26-ft. tides are not uncommon on this part of the coast - so we tied up to the piles below the jetty. Above us, into the darkness stretched two iron ladders. I was gingerly ascending the less rickety of the two, when a voice from above advised me to take the other one, as it was safer. Personally, I had doubts about both of them, however, we managed to get ourselves and our gear ashore, then bade farewell to our friendly tug-boat skipper.

We were now in the Indian village of Kitimat. It was a beautiful summer night, and we could smell the wild roses along the trail as we walked. Our guide took us to the house of one of the councillors, who in turn took us to the Chief's house where we were to be quartered. The Chief was away fishing, although we met him briefly later on. The house was the original Anglican Mission house, and was very comfortable; it even contained a piano, somewhat out of tune, but playable.

Across the Douglas Channel was being built the new town of Kitimat, with the aluminum smelter, townsite, wharfs - a complete new city being carved out of the wilderness. The sounds of large-scale construction drifted across to us in the daytime, and part of the night also. This project did not affect us, practically, in any way; except that we were able to buy groceries on that side of the channel which were not obtainable at our village store. In a way it was quite agreeable, after a visit in the councillor's boat to this hive of activity, to return to our quiet little village, and carry on with our work, unharrassed by the machinery of the twentieth century. There was no doubt that this village had prospered by being out of contact with the white man. The villagers, though without television, and other so-called trappings of civilization, were contented in their isolation; there had never been any crime in the community; and if they waited until the snow came, before getting in their winter's wood supply, that was the way of the native, who finds such behavior perfectly natural. A relaxed attitude implies neither a lack of enterprise, nor a lack of moral fibre, and - in this modern age - should logically be rated as a virtue. In our society nobody is content to merely live; we feel guilty unless we are planning, producing, performing or otherwise engaging in the business of simply being busy.

We spent about three weeks in this village, dividing their homesites into rectangular lots, and streets, according to instructions. There was no indoor plumbing in the village, and each lot had its privy at the back. I could have wished that they had not been quite so far back on the lots as they were; running the back boundaries proved to be quite an odoriferous business, with the lines running right through the privies.

We were very well received by the villagers. These people were, of course, much excited by the development taking place across the channel, and looking forward to the benefits that would accrue to them, due to the proximity of the new townsite. I felt that some of the blessings might turn out to be of the mixed variety.

A few days before we were finished, and bearing in mind the difficulties we had recently encountered in travelling, I contacted the Project Engineer's office at the new townsite, to find out if I could book seats on a plane to take us out. The answer was what I had expected - only V.I.P's could expect a firm booking, and we, who weren't even employed on the project, would be at the bottom of the waiting list. This time I didn't fool around. There was a seine-boat at the village, and it had just finished fishing. We chartered it.

At 6 o'clock in the morning on the day after we had finished our work, we were putting our gear aboard the "Matthews Bey", and with the whole native family who owned her as crew, we set off down the Douglas Channel en route to Prince Rupert. The fact that fog in the channel had reduced visibility to fifty feet did not dampen the enthusiasm of this crew. The engines ran at full speed ahead, and nobody seemed to worry. I must say I was quite relieved when the fog eventually lifted, and we found ourselves in the middle of the channel.

Two hours away from Prince Rupert, I called on the radio-phone to a hotel, which shall be nameless:

"This is the seine-boat, Matthews Bey. We will be in town in a couple of hours, and I'd like to book four rooms for the night".

The answer came back loud and clear:

"Who are you, native people or something? No, we don't have any rooms".

I admit that, having been familiar with the area for some time, I ought to have expected this, but I didn't. It made me feel if anything even more outraged. Especially since they professed to have no vacancy, so I could hardly tell them what to do with their rooms! It seemed that after having enjoyed the friendliness and hospitality of native people for 6 weeks, civilization was starting to show its more disagreeable side.

Fortunately there were other hotels in Prince Rupert which did not discriminate.

SMALL PARTNERSHIP.

*"Thrift may be the handmaid and nurse of
Enterprise. But equally she may not..."*

– John Maynard Keynes (1883–1946)

Many a practitioner will acknowledge that running a small business, while giving full rein to ones sense of independence, entails working long extra hours, and even foregoing holidays as well. This is a syndrome familiar to all individual operators of small enterprises.

I was fortunate in 1953 to meet with another newly qualified surveyor, who thought as I did, that a partnership might be instrumental in obviating some of these problems. Tony was 10 years younger than I, born in Wolverhampton, England, having arrived in Canada via Ontario's assisted immigration scheme in 1947. Articled to the same mentor as myself and qualified a year after me, he had an office in Whalley which we ran for a short time; then concentrated our efforts in North Vancouver. Tony and I got on well with each other, even though we were both "worry-warts". But that was to the good. (There is one of these animals lurking beneath the surface of almost all competent surveyors).

One advantage of partnership is that you can discuss technical problems with your partner in order to get a different viewpoint; yet since we both inherited the same principles from the same mentor, Tony and I seldom disagreed on a technical problem. Obviously, this helped to harmonize our relationship. Precision was always Tony's goal, and his bible was the old Official Surveys Act.

From the start of our partnership, which coincided with the start of the first post-war real estate boom in B.C., our main source of income was the town and suburban work. It became, as it still is, the bread and butter of most surveyors in private practice. Therefore, it was inevitable that the out-of-town jobs, which to me were always the more interesting ones, would decrease as time went on. City and suburban work is "run-of-the-mill" - it is every day work - so, of course, it becomes often routine. Although I could fill a book describing the surveys I have done in Greater Vancouver, the task of presenting them in an interesting fashion is hard. So, in these pages, there will only be brief mention of such work. However, despite the routine effect of the "sidewalk surveys", both of us found the day to day running of a small business far from routine, and a considerable challenge.

When the flow of work was more than steady, we worked long hours, and went out of our way to train assistants, hoping that by the time they were trained, we would be better able to cope with the increased work. What frequently happened was that as soon as we had adequately trained assistants on hand, the flow became less steady. Sometimes it stopped completely for short periods, and we found ourselves hard put to keep men employed. This feast or famine situation seemed to be inherent in small survey practices. Even the larger firms had such problems, but since larger firms tend to undertake larger, and longer lasting jobs, they are better able to handle their manpower by hiring for the duration of the particular job only. When a surveyor's practice consists of many small jobs, which must be handled as soon as requested, the sudden drying up of work can be frustrating. It was quite possible to find that the result of ones endeavors was merely going to pay the hired help, and nothing left over for oneself.

We eventually hired a secretary/book-keeper as office staff. She came in for three half-days per week. The other days, we were our own receptionists. I always looked forward to the days when Bodell, our first secretary, came in; as, for the whole day the office was transformed into something much more exotic than a mere business establishment. I still don't know what perfume she used, but it wafted tantalizingly into every corner of the office, and though it frequently took my mind off my work, I would have been most disappointed if she'd stopped using it. Do not imagine, however, that our office had wall-to-wall carpets, a well-appointed waiting room with easy chairs, magazines and vases of fresh flowers, and all the trappings of today's business premises. Bodell's perfume was the only concession to comfort in the place, which was essentially Spartan. There was not even a separate "lady's john" - today an unheard of state of affairs. But all was in keeping with our shared ideas, to provide for what was necessary only, since it all came out of earnings. There was no way that we would borrow money - a naive attitude, perhaps, and certainly not tax-wise by today's standards.

I had a great respect for my erstwhile partner who, after we had split up up some ten years later, went on to run is own firm for many years. We became very close in our private lives to the extent of him asking me to be best man at his wedding. I recall many outings he and I enjoyed, mostly to do with catching fish, and sailing.

In those days, Fisheries regulations were not as rigid as today. It was possible to leave the house for the Chehalis or Alouette River 6 a.m., catch a steelhead trout (if one was an expert as Tony was) and return in time to eat it for breakfast.

I do recall one occasion, however, when I insisted on bringing a Coleman stove plus bacon and eggs in the car, so as to eat breakfast before we actually fished. We parked by the bridge over the Chehalis just below the pool where the steelhead were wont to rest in the morning, after having spent the night coming up river. It was 6 a.m., and as we eat another car drove up, disgorging two fisherman, dressed in waders and with gear at the ready; who after saying 'good morning', proceeded up the trail.

Not ten minutes later they were back, each with a huge fish (probably over 20 lbs each); which we duly admired (although with clenched teeth). When they had left Tony said, "How dumb could we be. If it hadn't been for your stupid breakfast, those would have been our fish!"

All was not lost, however; the next pool - higher up the river - yielded us a fish each (but definitely smaller).

Sailing became Tony's hobby, at which he excelled, ending up in later years as Commodore of the West Vancouver Yacht Club. Racing, not cruising, was his forte, and he was very successful at it. He served on the Executive of the the West Vancouver Club, was elected Commodore in 1976, and in 1983 he was made an Honorary Life Member for his work on acquiring out-stations. In the fall of 2008, Tony became seriously ill with COPD, and he died on May 3rd to 2011.0.00cm

CANNON AND COW-FLOPS.

"Round many western islands have I been"

– John Keats (1795–1821).

May of 1953 was "Parenthood Month" for my wife and I. It was a girl, and we called her Wendy because it seemed that she had wings to fly. She was a beautiful baby. Up till that time, I had never cogitated much on the subject of babies. A baby was, to me, simply a baby. But this one was quite unique (she was ours), and obviously the most beautiful one ever born.

Unfortunately I would have to leave her later that summer, as we had undertaken some more work for the Department of Indian Affairs, this time in Masset, Queen Charlotte Islands. It was to be a three week job, and I was elected.

We needed an extra assistant for this work. At the time it was hard to find survey helpers with any experience. Most applicants for such a job had never worked on a survey before, but we considered one qualification essential - they must be used to "the bush". Our extra hand for this survey was chosen by the expedient of taking him out on a local survey, where he had to "cut line" all day in the pouring rain. At the end of the day, he still seemed cheerful, and remained cheerful when we told him that it would rain like that in Masset for the next full three weeks. So we hired him. The system worked well this time; our helper became quite competent; and it did pour with rain most of the time.

While on the subject of pouring, I should record that this was the method used to get me aboard the Union boat from Prince Rupert to Masset. I had left Vancouver two days ahead of the crew, who travelled by boat; flew to Prince Rupert, where I liaised with various government departments. After which, followed a joyful reunion with some old friends in the town. Crawford M. was one of these cronies, his favorite haunt being an upstairs bar known as the "27" club. Here we ended up an hour before the boat sailed, which was at midnight. I was very glad of Crawford's help in getting aboard. I assumed "my boys" were already on board (they were), and I retired to a cabin which seemed to be deep in the bowels of the ship. In my befuddled state, I remember thinking that this would be a good place to sleep, as it would be near the ship's centre of gravity.

As soon as we left harbor, the S.S.Cardina went out of her way to prove how fallacious was my theory. She put her bows into the wave, reared up, then flopped into the trough, at the same time managing to impart a small sideways motion into the overall up and down movement. It was the sideways wiggle that was my undoing. I am not a bad sailor. In the Sicilian landings, ten years before, where every L.S.T. was top-heavy with guns and tanks, and the screw came whirring high out of the water with each wave, I was one of the few "pongos" who turned up for meals in the wardroom. But Queen Charlotte Straits in a westerly, after a boozy evening, was too much for me; I was as sick as a dog.

At breakfast, after the weather had improved somewhat, my "crew" inquired after my health, admitted that it had been a bit rough, and assured me that *they* were just fine.

Masset was an interesting place. Old Masset was the First Nations village, extending along the shore of Masset Inlet, while New Masset was the name given to the small village which had grown up around the government dock, and the mouth of the small river, south of the Indian Reserve. New Masset boasted two or three shops, a garage and a "temperance hotel". The resident Indian Superintendent also lived in

New Masset. We stayed at the "temperance hotel", which was simply a boarding-house. I seem to remember that if one wanted a case of beer, arrangements had to be made for it to be flown in from Prince Rupert.

One of the interesting features of Masset in 1953 was a herd of cows, which spent their time wandering aimlessly through the streets, completely uncontrolled, eating anything they found in front of them, and fertilizing whatever happened to be behind them. These cows, unlike the ones found in India, were far from being considered sacred - the reverse, in fact. Everyone to whom I spoke seemed to think that they were a profane and blasphemous nuisance. In the 1930's, one of the missionary fathers, an enterprising man, had succeeded in inducing the natives of the Masset Reserve to keep a herd of dairy cows. For some years all went well, but when the good father departed from Masset, the cows became neglected, and semi-wild. Although the natives were obviously unwilling to look after the animals, they meticulously kept track of the ownership, and woe betide anyone who tampered with one. At the time when we arrived, there was a court case pending, concerning one frustrated white resident, who, tired of having cows eating his cabbages and defecating on his doorstep, shot one; and since it was inside his fence, he was claiming the meat as compensation for himself. I do not remember the outcome of the case. However, twenty years later, when I returned to Masset to do another survey, there were no cows left, and we were able to walk the streets unhindered by the necessity of stepping between cow-flops.

The name of the Chief of the Masset Band was Matthews. He was a very able man, and we had much respect for him. He was doing his best to bring modern-day thinking into the village, in spite of the paternalistic attitude of that branch of the Federal government for whom we were working. He had been to London, England, and met with Queen Elizabeth at the time of the coronation, and was very proud of this fact. He was also proud of his antecedents, and I remember several interesting discussions in which he explained the genealogy of a Haida family, as determined by the basic Eagle and Raven clans; this well-conceived custom, which allows for a male to be Chief, but which traces descent through the female lineage, is both simple and convincing as a means of determining succession.

The Haida peoples on these shores in 1953 were in a state of transition, and it has taken more than five subsequent decades for their history to become better known, and their culture widely accepted as it is now.

Two centuries ago, before the white man introduced them to those blessings of civilization which almost destroyed them, whiskey and small-pox, they were an unconquered nation. Along the beach at Old Masset in 1953, I counted four old rusting cannon; they were small guns of about a three-inch bore, possibly bow-chasers, and firing a solid ten-pound shot. Mr. Matthews told me who the owners were - all different families, one being his own; but none of the owners seemed concerned with preserving them, or putting them under cover.

The rest of the story I heard from the Indian Superintendent. They were old Spanish guns, captured from a man-of-war in the 1700s, when Perez and Quadra, amongst others, penetrated to 55 deg, N. and beyond; the B.C. Museum had at one time offered to put the guns on display, together with the authentic historical records, and giving all due credit to the Haida Band for the exhibition, but although the natives had at first been interested, they would not agree to release them. Now that there is a museum in Skidegate, it is to be hoped that the history of these guns will become apparent, and their owners' families mentioned.

On Sunday, we took time off from work, on the assurance by Mr. Matthews that, if we moored a boat "just off-shore from the Community Centre", and about a quarter-of-a-mile out in the inlet, we would catch a halibut. He was kind enough to lend us his own dug-out canoe - a heavy boat, with anchor and halibut line complete.

Masset Inlet is really a channel connecting a large inland lagoon with the sea outside. There is a twelve knot current in the channel most of the time - going in or out - with slack water of only half-an-hour between tides. According to my reading of the tide-tables, we had about five minutes in hand before slack water. However, after we had rowed offshore to the appointed place, it was obvious that we were still moving fairly fast up the inlet. So I let down the anchor, which was - I now have to admit - not a very smart thing to do. As soon as the anchor, which was in the form of a grapnel, caught, the boat turned into the current, and the bows all but submerged, while the incoming tide raced past making a wash like a power boat. To pull up the anchor was not possible, although I could have cut the rope. Instead, we sat and watched, while the wash slowly subsided.

We let our lines down. We did catch a halibut, which we had great difficulty getting into the boat. It was only forty-five pounds, a mere chicken, but it was a foot wider than our boat's beam, and no sooner had we wrestled the fish inboard, than we noticed that the boat had turned 180 degrees, and the wash had started on the bows again. It was time to go.

Naturally we gave half of this big fish to Mr Matthews (who had made the catching of it possible) - and the rest to our landlady, who strangely enough was disappointed not to get all of it. On the other hand she was able to feed us a fish diet for the rest of our stay.

Some days later, we finished our work, and travelled back to Vancouver via the plank road to Tlell, ferry across Skidegate Inlet, then plane to the south. Alec, who was our number one assistant, later to become a partner, had good reason for wanting to get back. He was to be married within the month.

I am reminded of another trip which I made with Alec as assistant, memorable because it demonstrated how violent the weather can be on the B.C. Coast in November. I have to admit that Alec seemed to take it all as part of the day's work - but then, he did spend time working on the Union boats, and had even been in the "wavy" Navy. But to me the sea was distinctly rough.

We left Vancouver by Union Steamship again. The SS Cardina took three days to get to Namu, where we surveyed a small foreshore lease, after which a seiner named the "Quathiaski" took us aboard to run us down to Jone's Cove, where we were to survey another such lease. Our plans, however, had not included the weather. Although at Namu all was fine, and the sea in the inside passage calm, as soon as we left the storm clouds blew up. Jone's Cove was on the mainland near Cape Caution (an appropriate name), and to get there it was necessary to cross Queen Charlotte Sound, a most unfriendly place.

For two days we stayed at anchor inside a sheltered bay known as the "hole-in-the-wall". Our skipper was a fisherman, and the crew was his fifteen-year-old son. We had plenty of grub, and a radio; but playing crib for two days - even with such variations as deuces and one-eyed jacks wild (and the astonishing scores which can result) - eventually palls, so next day at 6:30 am we poked our nose outside the bay. I was just starting to wake up at the time, but what I saw out of the port-hole completed the process very quickly. It was the sight of the waves breaking over some isolated rocks, causing spray which went right over the top of our masthead. There was some intricate manoeuvring to get the ship turned around - no easy matter - and then to our relief, we watched the rocks going by on the other side, as we fought our way back to the "hole-in-the-wall", where we again dropped anchor.

Next day the wind had abated a little, so we tried again, and this time did get going. Even so the seas were enormous. Alec made some coffee, and we ate some bread and jam; but for me it was a waste, as my breakfast went over the side five minutes after I had eaten it. As we neared Jone's Cove, and looked at the line of foaming breakers on either side of the entrance, I thought how lucky we were to have a good, solid 3-cylinder diesel engine, throbbing away inside our boat.

It was a *wet* survey. Although there were no breakers inside the cove, there were a lot of swells, which kept us soaked from the waist down all the time.

But it wasn't till the return journey, the storm having blown itself out, that we realized how well the Old Man of the Sea had been watching over us. This was when the bilge pump clogged, the water level rose until it reached the fly-wheel, which then sprayed foul water and oil all over the engine room, and our fifteen-year-old crewman, whose language was not at all appropriate to his years. Personally, I felt that thanking Providence would have been a more fitting reaction to the event. Had it happened thirty-six hours earlier, we would have been surveying Davy Jones' Locker rather than Jone's Cove.

SOME ANIMAL ENCOUNTERS.

"All animals are equal, but some animals are more equal than others".

– George Orwell (1903–1950).

The big event for us in 1954 was the arrival of David on December 13th, at 7 lbs 1 oz, with a fringe of red hair. The red hair was a bonus from Kitty's side of the family.

So now, where there had been two of us, there were four, and since Kitty was thirty-eight years old, we decided to rest on her laurels. We had planned our family affairs very logically, and it seemed that all plans had worked out well, in order of priority - for example, first a business then a permanent home, after which (but not before) a family would be permissible. Dame Fortune could not have treated us more kindly by arranging both a girl and a boy. Had our future lives continued in such a well-planned fashion, this story might well have turned out to be pretty dull. As it was, before six years had elapsed, we made a whole series of unplanned decisions - and life continued to be interesting.

1955 was also an interesting year - it was the year we made an excursion to the West Kootenays. The word "excursion" implies a pleasure trip, but this was definitely a working visit; although we all seemed to derive a certain amount of pleasure from it.

Two groups of Mineral Claims were to be surveyed for Crown Grant purposes. Both sets of claims were in the mountains above Camborne, on the Upper Arrow Lake, at elevations of approximately 4000 ft. and 8000 ft. respectively. Undersurface rights could still be Crown Granted at that time, and for this purpose our survey was a necessity.

In 1955, the Arrow Lakes were still undammed. Revelstoke was a small town, whose main reason for existence was the C.P.R. roundhouse and engineering shops. To get to Beaton, which is now under the lake, it was necessary to drive south from Revelstoke on a road which crossed the Columbia River three times by means of reaction-type ferries, to Arrowhead (now also under the lake), where a ferry made a scheduled run to lakeside ports three times per week. Beaton was one such port; it contained a few houses, a post-office-cum-store, and a hotel with beer-parlor. At Beaton two roads started - one to Camborne, the other over the hills to Ferguson and Trout Lake. These small places, consisting of maybe three inhabited buildings, and others uninhabited and ramshackle, were all that remained of thriving communities in the boom mining days of the early 1900's. At the time of the boom, Camborne had three hotels, and a population of 12000. In 1955, there were still people there, as the Sunshine Lardeau Mine had restarted operations; but this mine closed up some two years later, and Camborne's population thereupon dropped to single figures.

The syndicate for whom we were working had been employing part-time a local prospector, who lived at Beaton. Henry had spent many years underground, but latterly (he was over 60 then), had taken to prospecting. In the village of Beaton, he was the sub-Mining Recorder, postmaster, big-game guide and beer-parlor waiter. In the bush he was a most interesting companion, and a stalwart one. He was to be with us as cook and packer, as well as general guide, since he had been present at the staking of the mineral claims.

Ernie was the second member of our crew. He was, and had been, steadily employed by my partner and I, and I had known him for some two years before that. He was a 'natural' survey helper - he seemed to know instinctively what the next move would be; in addition he was very strong, good in the bush, and easy to get along with. Well-educated he was not, and it seemed unlikely that he would become more than a good assistant on a survey; but he was undoubtedly the best assistant I ever worked with, and I always felt completely confident on any job if he was on the crew. The other member of the crew was a well-recommended young student, experienced in the bush and an aspiring hunter.

In these days of helicopters and electronic distance meters, it is strange to look back and recall the amount of time and effort that used to be necessary to get a small party to an elevation of 5000 ft. in a remote area, and to keep them there for 3 or 4 weeks whilst getting work done. In 1955, we could at least get to the bottom of the hill by automobile, but from there on we were in much the same circumstances as, say, our predecessors the 1902 survey party. They might in fact have been better off than us, since they did have horses. In the manner of doing their calculations (by logarithms) we were no better off, since we had discarded our hand-cranked calculator as being too heavy for back-packing.

The three of us set off in my station-wagon from Vancouver, loaded down with equipment, bedrolls, two tents, food and personal gear. At Revelstoke, I visited the Mining Recorder, and we stayed the night in town. Next morning we drove to Arrowhead, where we left the car, and loaded all the gear on the ferry. At Beaton, Henry met us with transportation, and we again transhipped baggage, and spent the night at his house. Next day we packed up, bought last minute items from Beaton's only store, and drove to Camborne, to the end of the Pool Creek trail.

Here much organization took place. We divided all the gear into two equal loads for each of us. We would proceed as far as the first set of M.C.'s with one load, then return for the second. Henry's pack turned out to be far heavier than anyone else's, at his insistence. We had to hoist it on his back for him. After we had done so, he sauntered up the trail with a cigarette drooping nonchalantly from the corner of his mouth. Periodically - from around the next bend in the trail - would come coughing fits, as though he were on his last gasp. But when I expressed concern, Henry assured me that all was well. Just a touch of silicosis, he said, and nothing to worry about.

At the old Harvey M.C., there was a log cabin. The snow had pushed it sideways, but the roof and floor were in good shape, as was the stove.

So we decided to sleep two in the cabin (on old wire mattresses), while the other two slept in the small tent outside. Then we went down, and picked up the other load, leaving Henry to get going on the supper. Actually he had started cooking on that stove at lunchtime. It really had never occurred to me that anyone could whip up a batch of cookies for lunch, yet that is what he did. He then announced that what we needed was a griddle, and he had just the thing back at the house; and that next time he went down the trail to get grub, he would bring it up.

We started surveying next day, three of us on the line. Henry, after showing us where the location posts were, became cook and packer - which meant that every three days he would go back down the trail, and come up again with more food, or anything else we needed. I had been dubious about packing a 15 lb. griddle up the trail, but after tasting the hotcakes Henry made, I changed my mind. I found that I could eat 12 for breakfast without any trouble. I must add that this was not mere gluttony. We needed to generate a great deal of energy. The claims were on a very steep sidehill, draining into Pool Creek, completely timbered, and we had to cut every inch of the line with axes and machetes (no power saws here).

We stayed in the Harvey cabin about eight days. While in that cabin, we had our first animal encounter.

By the dawn of the first morning, it became obvious that we were not the only inhabitants of the cabin. While we occupied the main floor, the attic was home to a colony of pack-rats, who -deliberately, in my opinion - kept us awake by scampering backwards and forwards above the ceiling. Probably they felt that we were the intruders. Usually I had no objection to pack-rats. They are pleasant-looking little animals, with floppy ears and bushy tails. But that first night, when one started walking around on top of my sleeping bag, causing me to become suddenly and profanely awake, I changed my mind about them. In the next bed was Ernie - now also awake - and after a hurried consultation by flashlight, we agreed that should the so-and-so creature come back, we would fix him good. We turned over to sleep, but not for long. Ernie's whisper came:-

"He's on the table!", so I shone the light and there he was, mesmerized in the beam.

Then I noticed Ernie moving the muzzle of his 30-30 closer, and closer to the rat until it was only 12 inches away. Fortunately for the rat, at this point Ernie found that he had the safety catch on, and the resulting click of his releasing it was enough to break the spell. The rat was suddenly gone - never have I seen an animal move so fast - leaving us with our mouths agape. So - another consultation - and this time the gun was left with the safety catch off. It wasn't long before our furry foe was back, and the flashlight found him perched on the edge of the table. This time Ernie made no mistake. With the muzzle of the rifle 6 inches from our unfortunate visitor, he pulled the trigger, and this time the rat disappeared for good, and even faster. The sound of the shot in the confined space deafened us for a while, but the bullet went harmlessly into one of the log walls. We turned over to sleep, and this time stayed asleep. The rat colony, from that night, on never bothered us again. Obviously they had got our message.

Next morning, strangely enough, the two men in the tent outside appeared quite ignorant of the whole affair, although I could not see how they could have failed to hear the shot - and wonder. I found the whole incident hard to believe myself; but there was no denying the small pieces of rat embedded in the walls of the cabin.

After finishing the lower claims, we again took to the trail, backpacking up to the "Alma" cabin, on the old Alma M.C. at elev. 5500 ft. This cabin was in an idyllic spot, in a meadow thick with wild-flowers. Whiskey-jacks flew down to greet us, and two chipmunks dashed in and out of the wood-pile. There was no unwelcome night life in this cabin. Henry had looked after it properly. Mouse-proof tins held any staple foods that were stored there, and we slept uninterruptedly. We spent a "Sunday" here, just relaxing. I remember taking a large tin tub down to the creek, where I built a fire and heated it full of water. Sitting in that tub was almost like being in Heaven. We also washed our clothes, and they dried very soon in the hot sun.

Next day, we took light packs, and went off up the trail again. The upper claims were still 2500 ft. above us, and it was plain that we would need a closer camp-site to work from. At about 7200 ft., we found it. It was the only level spot on the mountain - where the original cabins for the old 1902 mineral claims had stood - dug out of the side of the hill. The remains of the cabins were there, vanquished by the snow, and flat on the ground. We moved enough of the rotten timbers aside to make a space for two tents, and then packed up the rest of our gear.

In 1902, an enterprising syndicate had mined low-grade copper up here, taking the ore out by pack-mule. There was a tunnel, said to be 500 feet long, which had been drilled and blasted using hand-steel. Our work took us up to a plateau, well above tree-line. The surface was largely bare rock, with snow patches, so we made good time, and were lucky enough to find two of the original corner posts of the 1902 claims.

Henry had been keeping an eye open for mountain goat. We decided that if any came near, we would take time out to shoot one, as we needed the meat. A trip down to Camborne now took two days. Not being a hunter, I had expected that shooting a goat would mean an exhausting, circuitous climb to get above the animal, followed by an equally exhaustive descent, packing the meat on one's back.

This goat-hunt turned out to be quite a different story. We woke one morning to find a whole herd of goats clustered on a small knoll, right beside the trail, not half a mile from camp. After a hurried conference, Del, who was the eager hunter, set off with his 30-30, after being exhorted by Henry to make sure that he got a young one. After about 45 minutes, we heard the shots, and the herd started to move off, leaving one yearling lying on the ground. Del had certainly bagged our dinner. Unfortunately, he had also hit the mother of the kid, who lingered just out of range, obviously wounded. There was nothing for it but for Del to go after her, and this he did; leaving us to butcher the kid. It took him all day to do it, but finally he finished the mother, who fell 300 feet down into a canyon. He was a very tired hunter when he got back to camp.

Next day, I stayed in camp and calculated, while the others went down and retrieved the old goat; which, after butchering, they hung in the mine tunnel. Those goats could not have chosen a more fortuitous time to visit us. The kid probably lasted us for two or three days, and by that time the job was almost completed. However, on the last day we thought we should try eating the old one, so Henry cut up the best parts, and started them boiling with a whole tin of Bulmer's dehydrated vegetables at about 2 pm. At about 7 pm, we felt hungry, so we served her up. Even the vegetables were rubbery.

Getting out of that camp was easy. We were able to go down with one large pack each. We were down in two days, with a stop-over at the "Alma" cabin. I am sure that we swam in the Arrow Lake when we got down, and I do remember sitting in the pub at Beaton, with an ice-cold beer in front of each of us. I also remember deliberately just looking at the beer on the table for about three minutes, in order to particularly savour the moment of downing it. It was absolutely and utterly delectable.

PARK BOUNDARY.

*"Here's a good rule of thumb:
Too clever is dumb".*

– Ogden Nash (1902–1971)

Henry was with us on another job which we did in the early days of our partnership. This work consisted of running part of the boundaries of Revelstoke National Park, our client being the Dept. of Energy, Mines & Resources, Ottawa. Since Revelstoke was near to Henry's home in Beaton, we asked him to help us on the job, and he was delighted.

These particular jobs - out-of-town, in steep and wild country - demanded a particular type of person as a survey crew member, one who in 1956 was often hard to find. The essential qualifications were that he must be physically fit and expertly bush-wise, and that he adjust well with the other crew members. Of course, a readiness to work ones head off whenever the occasion demands is a welcome trait in any crew member; and for the Park boundary survey we got three assistants who fulfilled all these conditions. It was a happy crew.

The third man, in addition to Ernie and Henry, was Izzie. Izzie was an Acadian, from Nova Scotia. Until we hired him, I myself had never met an Acadian, but afterwards came to the conclusion that if Izzie were a typical representative, they must be stalwart people. He had other qualities, mind you, besides being a tireless worker and good bushman.

He was a dedicated chaser of women, particularly after having had a few beers, when his hands itched for something to "paw". I admit that I quite sympathized with this attitude, having often been, at his age, a frustrated pawer. Maturity, of course, lessens the compulsion, and as one gets older one likes to admire, but hesitates to touch. However, no such scruples about age bothered Izzy; in fact, he liked to pick them quite young. We heard later, after he had left us, that there was a trail of angry fathers clear across Canada, who would have liked to get their paws on Izzy; and we thought it was another such episode that caused him to move on eventually.

Revelstoke National Park lay within an area known as the Railway Belt, the strip of land approximately 40 miles in width, along the route of the C.P.Railway which, under the terms of Confederation had been given up by British Columbia to the Dominion Government (it was transferred back in 1930). The area had been surveyed by the Government of Canada according to the 4th, system of survey for Dominion Lands i.e. in quadrilateral Townships, Ranges and Sections, with boundaries running due North, South, East and West, and no road allowances. The "no road-allowance" regulation - which determined the 4th system - was decided upon because of the mountainous terrain; nevertheless, although the Township system may have been ideal for surveying on the prairies, it took first prize for unsuitability when applied to a mountainous region, such as the Selkirk range of British Columbia. Very few of the Base lines, Township lines or Section lines were ever run on the ground; instead, ground survey consisted of a series of traverses up the more accessible valleys, setting monuments which were then, by calculation referred as witness corners to the various Section corners. The Section corners, therefore, were to all intents and purposes theoretical, most of them being inaccessible on the ground, as were many of the boundary lines which joined them; but they were faithfully plotted in their correct geographic location upon maps of the area.

And since the authority for their appearance as lines upon the official maps was the same authority that created them i.e. the Department of the Interior, then the boundary lines existed, although they had never been run on the ground. An odd situation, you may say; and yet a not uncommon one to meet when performing legal surveys. To proceed farther, it was only logical that the bureaucrats who were called upon to describe the boundaries of Revelstoke National Park, should do so by referring to certain Section and Township lines, which (if you are still with me) had never been run on the ground.

So there you have the reason for our survey. We were to run these lines, according to the Canada Lands Surveys Act, and flag them, so that the Park Wardens would (at long last) know where the Park began and ended. In point of fact, the Wardens had a rough idea of where the boundaries lay. The ones we were to survey lay somewhere on a steep sidehill between the top of the Park plateau (elev. 6500 feet.), and the Big Bend highway at the bottom of the mountain; an area unmarred by road or campsite, with an average slope of 30 degrees, drained by numerous steep-sided creeks.

The working day began at 8 am., on the Big Bend highway, at a location vertically below the last station on the previous day's survey, and with an hour and a half's climb to get to our start point. On the first few days, it was 9.30 am by the time we had reached this point where we were to start the day's travail; but as we progressed, getting further from, and higher than the highway, the climb up took longer, till at the end of the job we were spending 3 hours climbing up, in order to work a 4 1/2 hour day. Coming down was a different matter - a half hour at the most, spurred on by the prospect of a cold beer. We were very fit physically when the work was finished.

The job was enlivened somewhat by a "spot of bother" that overtook Henry one evening. It was, of course, normal procedure for us to "wet our whistles" with a few beers after supper, but mindful of the hill-climbing next day, it was never more than one or two. On this occasion I had gone home, leaving Henry with one of his cronies at the hotel.

Now Henry, through years of packing heavy packs in the mountains, had developed an unusual posture. When standing, he had a pronounced forward lean from the waist. Moreover, if he put his hands in his pockets whilst standing, and one viewed him from behind - well, one cannot really blame the young constable, who did not know Henry's stance, and thought he was relieving himself. Henry insisted that he was simply standing in the alley, waiting for Izzie to come out of the pub. (We believed Henry - who, when we saw him in the pokey, was not drunk at all, but feeling a subdued sense of outrage).

It would not be unusual for the police to patrol the lane behind the pub looking for trouble-makers, but that this should happen to an amiable guy like Henry was quite unjustifiable.

He spent the night in the cooler, and though I interviewed the Mounties, the magistrate and the Prosecutor (another Mountie), and told the court what an excellent fellow Henry was, he was still charged with *"indecent exposure"* and fined fifty dollars for, as far as we could see, the crime of being in the wrong place at the wrong time.

I did not mention this incident in my Chief-of-party's report, as part of the Canada Lands Act survey, but later I wished that instead of being trained as a surveyor I had been a lawyer. After all - if *"exposure"* is a necessary act, how can it be *indecent*; and what is the meaning of the word *"indecent"* anyway . . . . yes, if we had had a smart lawyer I think we could have got Henry off the hook.

But it would have cost a lot more than fifty dollars.

BOOM YEARS AND A BUST.

> *"I can't explain myself, I'm afraid sir,"* said Alice, *"because I'm not myself, you see." "I don't see,"* said the Caterpillar.
>
> – Lewis Carroll (1832–1898).

For most British Columbians the years 1957 to 1959 were boom years.

Our family had moved into a larger house on upper St. Andrews Avenue in North Vancouver. It was at the end of the road, where a small stump-farm blocked further development. Our neighbors, who owned this one-cow farm, eventually sold it to a developer, and now St. Andrews continues up the hill for half a mile further. It was a nice neighborhood, far enough away from the city to be peaceful, yet only minutes away from our office.

We certainly crammed a lot of leisure activities into those years. Kitty and I made our first camping trip into the Black Tusk meadows of Garibaldi Park, while a kindly Norwegian lady looked after Wendy and David. Then, with the kids, we spent a week on sand-girt Savary Island, where we lazed on the beach. Now we understood why families with small children always opt for holidays at the seaside; all one has to do is turn the little tads out on the sand, and you won't hear a peep out of them for hours. If only Savary Island were surrounded with *warm* water

. . . . .

But to catalogue our family activities would be quite beyond the scope of this narrative. Moreover, the good years were deteriorating. The year of our last holiday was 1960 - a bad year - and by that time we had decided to leave British Columbia for warmer climes - New Zealand first, then maybe Australia. So if the following pages read like a travelogue, it is because we covered a great deal of land and sea in the next twelve months.

Looking back at this upheaval - for upheaval it was - it is hard to avoid the assumption that we were both at heart 'rolling stones'. Perhaps making a living had been too easy (a most incorrect assumption - it hadn't been). Naive we certainly were in spite of our ages (I was 43 years old), as we realized when we returned later to Canada. Opportunity maybe knocks once, but it doesn't make a habit of it, and what we failed to realize was how well-off we were in spite of the recession of 1960. The "boom and bust" cycle of North America was a phenomenon we knew little about, but had we known more, we should have realized that the economy would eventually upswing again. Other considerations were:- "There might be another war" - (the Bay of Pigs episode was still current news at the time) - the warmer climate down south with its swimming, tennis, sailing and hiking - the subordination of Canada's culture to that of the U.S.A. (had we been residents for longer, we would have found that this had been an ongoing process for many decades, and that Canada was well able to hold its own) - and add to these considerations the wishful idea that we might be making a very smart move, Canada's post-war boom being over, but Australia's and New Zealand's about to start.

I am sure that a two month's holiday from work would have cured us of the wanderlust. From the time my partner and I started together we had never had more than an annual three week's holiday, and in the winter-time at that. Many people in a position similar to ours get "itchy" feet, and a long holiday is probably the best cure. But - regretfully - such measures were not for us; we *liked* to do things the hard way.

It is safe to say that all our friends advised us not to go. But when they found that we really meant business, they held "going-away" parties, to wish us luck. There were more than a few of these parties. Granma Bakewell, for her part (and she knew what she was talking about), warned us that Australia was "full of fleas" (we found this to be true); and while the grown-ups were socializing at her house, David complicated matters by breaking his arm. At the time, he was thought to have fallen out of an apple-tree (normal boy-like behavior). We know now that he was really jumping off the roof of the car, which (although just as boy-like) would probably have been viewed with less tolerance, had we known the truth. So he accompanied us overseas with a plaster cast on his arm. This impeded his activities hardly at all, in fact he found it quite a useful weapon when skirmishing with other kids.

There were "family" in New Zealand. Cousins Jim and Pauline Cotton lived there, and sent us much helpful advice. But I will omit details of the very extensive arrangements which we found necessary in order to do what we did. Naive though we might have been, we were remarkably efficient in all our organization; not that this helped much in our dealings with the many bureaucracies with whom we came in contact, all of whom were quite the reverse. We sold house, cars, cashed-in our retirement savings plan (how provident I had been up till that time). The only property we didn't sell was at Tyaughton Lake. Dave, our partner in that enterprise, simply said: "There's no need. You'll be back!"

Our passage on the "Canberra" had been confirmed the day after we sold our house, and a few days before sailing, we moved into a motel on the Capilano Road, all our belongings save hand-baggage having been shipped; there were 21 numbered pieces of freight. Thence to the ship, which we boarded in the afternoon, and - surprise - there was a party in our cabin, organized by my ex-partner and "the boys". Some other people showed up too, notably Crawford M, and we consumed several bottles of pink champagne.

And later, after the farewells, and the streamers, we sailed under the Lion's Gate Bridge, and watched the lights of West Vancouver recede, holding two sleepy kids, who really didn't know what it was all about.

Any more than we did.

Angus Beaton and pack horses.

B.C./ALBERTA BOUNDARY SURVEY 1950.

Bob Thistlethwaite at witness corner.

Clare Hobbs at Lake Lovely Water - 1959.

BRITISH COLUMBIA.

Dave Underhill in the Monashees - 1951.

Stump set-up.

Hamish and I - Skeena River 1950.

D.R.Bakewell at Glacial Creek - 1946

BRITISH COLUMBIA

J.T.Underhill & helpers 1947.

The "Please Go Easy" railway.
- Marne 1947.

Alec in 1953, Masset, Q.C.A

Sakinaw Lake - 1958.

BRITISH COLUMBIA.

A glimpse of Tyaughton Lake - 1960.

Pt Atkinson - 1958

David - 1960.

*"Girl and dog"
Wendy with
Rusty - 1958.*

Rolfe & Bill.

Mineral Claim Post.

PIEBITER CREEK 1951.

The Office.

The Cook.

"Back to civilization"

"A bunch of the boys" - Mark, Harry, Keith and H.B.C.

Instrumentman.

Fording the Kloia River - Keith & Harry. (It was a hot day).

Woodstave pipe.

PRINCE RUPERT.

Columbia Cellulose Surveys 1949 - 1950.

J.W.Hermon
(Hamish).

H.B.Cotton
(Barry)

A.C. St P.
Bunbury.
(Alex)

Tim, Ralph & H.B.C.
(Obviously posed) 1962.

1965 Partnership.

TYAX DAYS.

Proof that I am not a liar!.

The woodchopper.

S.M.H.A. Bldg - Cooma North.

Tumut Pond Dam.

Flags of all nations - Cooma.

AUSTRALIA.

Lake Eucumbene.

We had to leave the kittens behind

Cruise ship.

Boiling Lake - Waimangu.

DOWN UNDER.

Lady Knox geyser (private viewing).

Nick

Ken & Flo Thompson

In 1967   *Wendy & David.*   *In 1987.*

RECENT DAYS.

Kitty in 1989.   *The author in 1987.*

5. DOWN UNDER.

TO THE SOUTH PACIFIC.

"He is no wise man that will quit a certainty for an uncertainty".

– Samuel Johnson (1709–1784).

There is no doubt that the three-week maiden voyage of the S.S. Canberra provided us with the best holiday we had had in years. It was three weeks of cruising on a luxury liner, even though we did have one-way tickets.

For that matter, our children did not suffer any hardship either. The concept of the English Nanny was, of course, new to them, and might have been viewed with some suspicion, but there were large, well-equipped nurseries on the top deck (well-staffed with the afore-said Nannies), and they spent a good deal of their time with the many other kids of the same ages.

That the British Empire was alive and well became apparent early in the voyage; at 7 a.m. on the first morning in fact - when, after a light tap on the cabin door, an immaculate steward was revealed (he was "Goyan"), holding a tray of early morning tea, which he put on the fold-down table, and proceeded to pour for us while we were still in bed. We did not object. However, it is worth noting that, although we were still in American waters for the next week, we felt increasingly that we were now in the sphere of British influence, an idea that was well confirmed during our stay in New Zealand. It was somewhat different in Australia, where the independent Aussies regarded both North Americans and British as foreigners. (We ourselves were never referred to as either "pommies" or "Yanks" - I don't think anyone really knew what to make of us).

Kitty and I were both seasoned steamship travellers, but our experience had been in war-time. This ship had all the facilities of a modern cruise ship; even in tourist class there were three lounges, dining rooms, ball-room, billiard-room, exercise rooms, saunas and half-a-dozen bars; and swimming pools; and orchestras playing every evening . . . As for the prices of liquor, at that time if a ship was beyond the three mile limit, alcoholic drinks were exempt from excise tax, and one of my fond memories is of paying no more than 20 cents for two gin-and-tonics.

At San Francisco we were welcomed by two fireboats, scores of yachts, towboats with flags flying and sirens blowing, as we sailed under the Golden Gate Bridge. At Long Beach, there was Disneyland, for two exhausting days. So when we got to Hawaii, we were looking forward to a little relaxing - Polynesian style - and what could be more relaxing than renting a jeep with a surrey on top, and fitting into it ourselves (two adults and 2 kids) and two friends we had met on board, (with their two kids), and driving around Waikiki. Unfortunately we soon found ourselves on the Pearl Harbour freeway, with all six lanes travelling at least 80 m.p.h. In those days rented cars had a governor, and our jeep, at 30 m.p.h. started to generate a lot of congestion, - not to mention horn honking - especially as we hadn't enough speed to change lanes to even get off the freeway. This was relaxation a la Hawaii?

Maybe not, but even so, we had seen enough of this beautiful land to want to return (even if we had to drive across the median to get back to our ship), and we made sure to drop our leis over the side as we sailed away (we *did* come back).

'Crossing the line' in a ship used to be a significant event, attended by mock-serious ceremonies (and practical jokes played on those who had not crossed the equator before). Today, when millions of people fly across without even knowing, the tradition must have almost died out. However, we had a great time on the Canberra's maiden voyage; with festivities planned mainly for the youngsters. As the 'doctor', I remember squeals of delight from the kids at the prospect of having their heads chopped off with a cardboard knife; and although the swimming pool was quite small, several adults ended up in it fully dressed, as expected.

Approaching New Zealand, however, two problems became evident. The day on which we were due to arrive in Wellington was the same day an international rugby match was to take place between New Zealand and France, also in Wellington. Now, although most New Zealanders profess to acknowledge Christianity as their religion, rugby football is the real shrine at which they worship. And it soon became apparent that there would be no accommodation whatsoever in the city that evening when we were due to arrive, as the whole nation would be converging on Wellington, both to see the game, *and* to celebrate the victory afterwards (there was no doubt in New Zealand as to which side was going to win).

In the event, the second problem solved the first. The second problem was a storm of nearly hurricane force, which arrived just before we got to Wellington. The swell was generating 26 ft. waves and to brave Wellington Heads in such conditions was not considered prudent. So, instead of entering the channel, we cruised up and down Cooke Strait, a waterway not noted for calmness, until the storm blew itself out. This took three days. By that time the football game was over, and all the fans had returned home. We docked at Wellington harbor with only a few smashed windows on the upper deck as casualties.

As we said good-bye to our shipboard friends at the Wellington dockside, we reflected that, while all this had been most enjoyable, the real adventure was now about to begin. Not that we felt any trepidation; it would all be a challenge.

And a challenge it was.

NEW ZEALAND I.

"Better ones own path, though imperfect, than that of another well made."

– Bhagavad Gita (c - 200 BC).

The best hotel in Wellington at that time was the Waterloo. In it, we had grandly reserved a suite on the assumption that, since the cost of living in New Zealand was lower than in Canada, the hotel costs would also be correspondingly lower, a quite erroneous deduction.

After having had a meal before unpacking, and then viewing the tariff posted on the wall of our room (accompanied by a dozen or so regulations - "by order of the management"), we came to the conclusion that, while the state of Denmark might not necessarily be rotten, something in New Zealand definitely was. Here was a hotel in a country with a lower standard of living than that which pertained in Canada, yet charging prices far higher than the Vancouver Hotel!

We decided to leave, and leave we did (to the management's consternation); putting ourselves in the hands of a sympathetic taxi driver, who found us a large double room in what seemed in the dark to be a temperance hotel. In the morning light, we found that it resembled a boarding house, remniscent of England, and was complete with dining room, in which little old ladies politely sniffed at the food and each other, whilst declining any conversation beyond: "Good morning", or "Pass the salt".

But the price was what I figured we should be paying in New Zealand, and that explained very well the lower standard of living.

But all things come to him who perseveres. The next cab-driver found us a motel - the first of its kind in Wellington. It had a kitchen, was modern, and the rent was reasonable. So there we stayed until my business in the city was concluded, and we could move on.

My memories of Wellington are of drab, grey streets, in which huge, slow behemoths of street-cars relentlessly moved, and where it rained incessantly. I found that I could usually walk faster than the street-cars, but it was hard dodging the rain.

My cousin, Jim Cotton, a medical doctor, had left the country some months before with his family, to go to Britain for post-graduate study. However, his parents-in-law, Ken and Flo, were waiting for us when we got off the train in Auckland, and it was a welcome which I'll never forget. This kind couple, who had never seen us before, took us to their home in Ellerslie, where, in a flower-girt orchard they fed us strange delicious foods, such as tree-tomatoes and Chinese gooseberries (now universally known as Kiwi-fruit (with appropriately large helpings of whipped cream) - drove us around Auckland, and took us to the beach. Next day, they drove us to a little cottage on Ellengowan Road, near Brown's Bay. They emphasized that the cottage was being lent to us, and would not even talk about rent. The beach was half a block away, with a small store-cum-post office (we paid a nostalgic visit to it again in 1983, and it had hardly changed - and nor had Ken, an alert 82, still enjoying life to the full).

Whilst at Torbay, I took the opportunity of visiting various survey offices in the Auckland district, travelling by bus; and in doing so, found that, as a qualified surveyor from another country, I was something of an oddity. Most assistants in the employ of surveyors were articled students, and it was easy to figure out that such employees, mostly unmarried, would not earn a great deal of pay. As for me, although experienced, I was new to the country and its methods, and yet it was plain that I would expect more pay than an articled student.

In 1960, New Zealand still had an adverse balance of trade with both Britain and the United States. There were restrictions on how much currency (sterling or dollars) could be taken out of the country. For this reason we did not pay our bank draft into a New Zealand bank, but kept it in a safety-deposit box (in case we should find that we wanted to leave in the not too distant future). Although as an immigrant, I was entitled to "bring in" an automobile, I would not be allowed to sell it freely. All cars in New Zealand had had to be imported, with a resultant outflow of sterling or dollars; and since the government took a dim view of the outflow, there were rigid import restrictions regarding cars.

From the first day in this pleasant land, I had viewed with wonderment the assortment of old cars that cluttered the highways. I never once saw a new car; but I often saw model A Fords; and once I saw a bull-nosed Morris Oxford (vintage 1928); some old cars were in excellent condition; others were held together with baling wire. The net result of the government's regulation seemed to be that, while a new car had a fixed value, as soon as it became second-hand its value could (and did) soar. A logical situation, but puzzling for outsiders to digest at first.

I picked up our car in Wellington, and drove it back to Auckland via the west coast road. On arrival at Ellengowan Road, I found that a job for me was in the offing. I would be working out of Harrison & Grierson's office at Rotorua. However, though a job might be available, accommodation might not necessarily be so. So I drove to Rotorua, and spent the night in what was definitely not the best hotel. The bed sagged in the middle, almost to the floor; and I was an object of extreme suspicion to the other residents, many of whom looked as if the last stimulating experience they had had was when Noah grounded the Ark on Mount Ararat.

Rotorua in 1960 was not impressive. There was no Sheraton, no International Plaza, no shopping centres and few restaurants. Whakarewarewa (Whaka for short) was there, of course, with its geysers, and very excellent Maori tour guides; there were a few gift shops, and the Municipal Spa, and the smell of sulphur was just as strong then as it is now. I drove a few miles out of town, to Ngongotaha on the lakeshore, where the air was fresh, and stayed the night at a motel. The family who owned it were most interested in our circumstances, and happy to accommodate us. I looked no further. Next day I travelled back to Auckland, to say "au-revoir" to Ken and Flo, and bring the family to Ngongotaha.

I went to work on the Monday.

NEW ZEALAND 2

"There is a lot to be said for challenging fate, instead of ducking behind it."

– D.H. Lawrence (1885–1930)

New Zealand's system of measurement was both modern, in that they used steel tapes, and old-fashioned, in that the tapes were graduated in chains and links. One needed a "link-stick" in ones hip-pocket, to measure fractions, and a spring balance.

Most of the other employees in the firm were articled students, for whom, I suspected, parents may have paid premiums as a condition of their signing articles. The work that I soon found myself doing did not differ materially from similar work in Canada - subdivision, and some engineering surveys; but generally I felt that I had time to work on a problem, and that there was noticeably less pressure than I had felt before.

Kitty lost no time in looking for a place that we could "call our own". She found a neat little summer cottage on the shore of Lake Rotorua, which could be rented for 12 pounds a week; we moved in. It proved to be quite cold, but it was accommodation, and we were lucky to have found it. There would be a future problem, however, in that, as soon as the summer season started, in December, the rent would go up to 25 pounds per week.

How much of a problem this could be was apparent when I brought home my first pay cheque, which was for 15 pounds; this didn't leave much over for other necessities beyond rent. I figured at the time, that my pay (for doing what any registered New Zealand surveyor would do, although I could not sign a plan) was about equal to what, in Canada, I would pay a rodman. Still, we had a house, and I had a job; and all acquired within the space of three weeks; so we hadn't performed too badly.

New Zealand is a delightful and beautiful country. We lived on the sandy shore of Lake Rotorua, where every beach and headland has its place in Maori legend. At week-ends, we took picnics and explored the thermal areas - Waimangu, with its hot running river - Waiotapu (the sacred waters), with the "champagne" pool - and the Lady Knox geyser, where we asked permission of the owner of the property to view, and she responded (on payment of sixpence) by wrapping the cone with a sack, and inserting a piece of laundry soap in to the vent; we watched in awe as the geyser materialized, eventually blasting forth with full power just for our eyes alone. At Lake Taupo we swam; and at Wairakei (which the Maoris called "the place where pools were used as mirrors") we toured the thermal power stations. We drove to Tauronga, and the beaches at Mt Maunganui, where, although it was cold, we swam in front of sand dunes bordered by miles of yellow broom. And everywhere there were green fields, and sheep, or forests of pine, with trees marching in perfect alignment up the hills.

It was a relaxed and rural country. We explored Rotorua too, Rainbow Falls, Whaka and the Maori "pa's"; and several times visited the local spa, to "take the waters". Lake Rotorua itself had to be explored by boat.

It turned out that one of our neighbors did have a boat. Also a daughter of Wendy's age. We went on several excursions, and then settled more or less into a routine in which - if our work schedules tallied - our neighbor and I would arrive home from work, jump into the boat (which was known as "Pops"), and spend the next hour in fishing. We never failed to catch our supper. Wendy and David, in company with a flock of children of their age, in bare feet - they never needed shoes - played in and out of the water. Their friends included a little Maori boy who went by the name of "Lucky Duck" - we never did find out his real name.

Yes, it was a delightful country.

When time came for school, our children were enrolled in the local school at Ngongotaha. The grade system was the same as in Great Britain. It was a bit unsettling for Wendy, who already had one grade in Canada, under far more progressive conditions for the teaching of early grades. But she assimilated well, even to the point of adopting a reasonable facsimile of a New Zealand accent.

About this time, it began to dawn on us that in less than two months our rent was going to double, and it was literally impossible to find another place to live. Also, although I had heard from the Survey Board quite favorably, it would be another two years before I could qualify in this country.

I wrote to the Snowy Mountains Hydro-Electric Authority, in Cooma North, Australia, listing my qualifications, and awaited a reply. It came promptly (surprise!). I was to go to Nelson, in the South Island, where I would be interviewed by an engineer of the Authority, after which I would be further advised.

I flew on a week-end, by D.C.3., to Nelson - then an unusual way to travel in New Zealand. In Nelson I had a successful interview. Three weeks later I was offered a job as Survey Officer Grade III with the Authority, commencing January 1st. This was something to think about, especially as the salary was about three times what I was presently receiving. But unfortunately the only housing available in Cooma North was for single employees.

But before deciding one way or the other, I was to become even more acquainted with New Zealand by going on a hunting trip. At that time - I preface a great many statements with this phrase, because now, fifty years later, circumstances are a lot different - there were no restrictions on hunting wild deer or pigs. Both species had been introduced many years ago by Captain Cook as domestic animals for use by the Maoris. However, due to the fact that there are no natural predators in the country, the wild deer and pig population in the back areas had increased enormously. Both animals caused so much erosion and damage to the environment, which the government intended to keep beneficial for the farmer, that hunting was encouraged in every way, without the necessity for even a licence; New Zealanders considered it their right to hunt freely any time they chose.

I owned a .303 Lee-Enfield sporting rifle, and it was not long before two local lads, both avid hunters, presented themselves, and offered to take me out. The reason was also apparent - I had a car.

It was to be a long day. At 3:30 am we drove into the hills south-east of Lake Taupo, following logging roads in original forest, till after about 25 miles from the highway we came to a shack. This shack - 12 ft. by 8 ft. - had been the abode of a would-be settler in the 1930s. After building the shack to live in, and clearing by hand about 40 acres, planting grass, and bringing in a herd of sheep (no mean accomplishment), he had then said: "to hell with it!", or words to that effect, and had walked out, abandoning the whole project, sheep and all. The sheep had now become completely wild.

Due to the clearing, this should have been a good place to find deer; but on this day the wind was gusty, and no deer were apparent. We split up, agreeing to meet at the shack later. After spending several hours bush-whacking, I headed back; and there, in front of me, was the herd of sheep, on my windward side. Any qualms I had about shooting a sheep disappeared as fast as the sheep themselves were about to. I nailed a yearling, and the sound of the shot brought up the other hunters at the double.

Altogether, we brought back a deer and a lamb, so it was a successful hunt, although it was 9 o'clock pm before we arrived home to share out the meat. It was some years later before I divulged to Wendy that I had shot a lamb. Had I told her then, her reaction would have made me feel even more of a criminal; so we said nothing, and simply ate the proceeds of the crime. It was delicious.

Six weeks before Christmas, we made up our minds that we would have to leave New Zealand, and move to Australia. I couldn't see us ever being able to "make ends meet" with the housing situation the way it was. My work was satisfactory, and I enjoyed it; the kids were happy at school; this country of New Zealand was a delightful place. But we were being defeated by the lack of accommodation.

So, after accepting the job offer at the S.M.H.A., and giving notice to my present employers, all the necessary scheming and arranging for a major move had to be started again. By this time we were becoming quite expert at the business. To make matters easier, our 21 packages of baggage were still in storage at the docks in Wellington (where, according to the cynics, most of it usually stayed till picked up again on the outward journey).

We bought international bonds with our New Zealand money; we sold our little red Hillman car (for the same amount that we paid for it). But we missed the obvious loophole - had we sold it to an intermediary, who could then resell it (with no restrictions on the sale, as it would now be a used car), we could have split the profits on the second sale - this we found out later.

This time there was no party in the cabin. A few friends saw us off on the bus; we stayed overnight in Wellington, but at the Salvation Army Hostel; the S.S.Aorangi, a Dutch ship, was not nearly so luxurious as the S.S.Canberra, and she didn't have stabilizers; we all got a little seasick before we docked in Sidney.

AUSTRALIA 1.

"Life is just a bowl of pits!"

– Rodney Dangerfield (1922–2004).

The S.S.Aorangi rolled its way across the Tasman sea, and put in at Sidney for two days.

It was not as though we weren't welcomed to the new country. Australian hospitality was almost overpowering. At Sidney, friends from our Canberra voyage met us, crammed two families into a car designed for four people, and gave us the grand tour. When the ship docked at Melbourne, two friends whom we had known in Vancouver were there, and helped us find our feet. To Mike this was a far superior country to Canada. I was not prepared to concede as much, but figured we would give Australia "a go" and see what transpired.

What transpired, however, was not encouraging. Although wages were higher than in New Zealand, and although I had a job with the Snowy Mountains Hydro-Electric Authority, there was still a housing shortage, and the problem of where to live as a family was unresolved. Moreover the place which we ended up renting verged on being squalid. That evening our spirits reached their lowest ebb.

Something had to be done about it, and Kitty, by working through the "want-ads" in the newspaper, found us a better place to stay. It was with a family on the outskirts of Melbourne, who had two girls of the same ages as our children.

My foremost memory of Melbourne was *not* nostalgic - it was of not being able to park any closer to the beach than eight blocks, finding the sea-water to be actually tepid, and then burning my hand on the side of the vehicle when opening the door. (We had acquired a used Volkswagen bus soon after arrival, as we still had the 21 packages to move around).

The Australian highway code then was a unique system. While driving on the left, the guiding principle was that the driver on ones right has the right-of-way, and one must give way to him. This could give rise to all manner of queer situations, which although normal to Australians, caused outsiders to drive with one foot on the brake, and a finger poised rigidly over the horn button. A couple of miles from the city centre in Melbourne was an intersection where five roads met. There were no traffic lights, or "stop" signs; and the traffic was intense. We called it "the chaos". When threading ones way across this maelstrom, one looked simply at the driver in the car on ones right hand side. If his lips were moving, he was either swearing at you, or praying; one never had to look at the driver on ones left, as he had always (you hope) to yield the right-of-way. A stressful situation - except to Australians, who looked on it as normal procedure.

Once out of Melbourne, and on the way to the Snowy Mountains, I was able to enjoy the country. It was not green. Near the foothills of the Australian Alps, there was some timber, and even small creeks. Yet there were few inviting places to stop, and I pressed on till night fell. At about eleven pm, I arrived in Nimmitabel, which had a wide street, and several hotels. A notice-board hung outside one of them, telling all and sundry that this town, and this hotel in particular, had been featured in the Hollywood movie, the "Sundowners", starring Deborah Kerr, and Robert Mitchum. It seemed a good omen, and I was hungry; so I knocked on the door, which promptly opened, and as I entered, a schooner (which is a large tankard) of beer was thrust into my hand. Events happened quite swiftly after that - it seemed that there was a wedding party in progress, and I was invited; introduced to the bride and groom, whose health I drank several times, and to many of the guests, who insisted on refilling the tankard at intervals.

Eventually, the landlord introduced me to a plate containing a substantial meal of steak and eggs. The party went on, in spite of a notice behind the bar, stating that closing time was 10.30 pm. At about 1.30 am, the landlord showed me to a room with a brass bedstead, old-fashioned dresser, and wash-basin containing a china bowl and jug of cold water. The toilet was "out-back", but I managed to find my way without hazard. It occurred to me that this was the real Australia, and once again, how could one fail to be impressed by such hospitality?

Next morning, more steak and eggs, and I went on my way, arriving in Cooma North, headquarters of the S.M.H.A. at about 11 am, where I was assigned a comfortable but cheerless room in a bunkhouse.

When the Authority had first been set up, one big problem had been how to arrange accommodation for thousands of workers, who would be spending their time building, engineering, surveying, drilling and so on, in a country that up till then had literally been the "outback"; where roads were only tracks, and villages few and far between. The Authority solved one of these problems by establishing "hostels" at selected places.

Administration generally seemed to have been modelled on that of the armed forces; e.g. - there were "officer employees", and "enlisted men" employees. The former had signed contracts with the Authority, the latter were simply hired. There were even security guards who could only have been patterned after the military police. In the matter of accommodation, the officer's quarters, and meals were on a superior level. I was to spend a good deal of time in such accommodation, but found that, after dinner had been served, the best entertainment (and the best beer) was to be found in the "enlisted men's" canteen, rather than in the "officer's mess". All such accommodation and meals were, of course, free for the user, whose only reason for being there was to work for the Authority.

Each new employee spent a few days in "orientation". This included two days on a guided tour of the various projects in the overall scheme (embracing 3000 square miles), in company with real tourists, who had paid for the privilege. Thus, at the beginning, one learned a great deal about the scheme as a whole, before becoming involved in ones own particular sideline.

It was an immense development. Its function was to dam the Snowy River, which yearly took all the run-off from the Snowy Mountains straight down to the sea, reverse its flow through various tunnels, till the waters descended into the Murray and Murrumbidgee Rivers. On the way down, the water generated power through six large and several smaller power stations, to supply the growing cities of Sidney and Melbourne; and was later used for irrigation - one answer to the problem Australia had had for so long - a simple lack of water. This problem becomes even more apparent, when one considers the size of the Snowy River; it was, in fact very little bigger than Mosquito Creek in North Vancouver.

There were altogether 100 miles of diversion tunnels, 80 miles of aqueducts and no less than 18 storage dams, the largest of which - Eucumbene Dam - was an earth-fill dam 381 feet high, forming a reservoir of 3.8 million acre feet capacity. In order to accomplish all this construction in the planned 30 years, the Authority had offered high salaries, and fares paid from their country of origin, to many of the best hydro-electric engineers in the world, since Australia's own engineers were too few in number to handle the development. Funds were provided by Federal and State governments each year, and in that context, there was a very excellent Public Relations Department, with tour guides, lecturers and press agents, to acquaint the general public with the importance of the scheme and its progress; thus it was very unlikely that any political party would risk the outcry that would follow an attempt to curtail funding.

The Survey Department in this engineering world was subdivided into several sections: e.g. - Control Section, Underground Section, Construction Section, Legal Section, and Topographic Section; the overall chief being a German Engineer of high qualifications. The first and last Sections were largely concerned with mapping; the second and third with purely engineering principles - laying out and aligning. The Legal Section, for which I opted (on the assumption that maybe I might aspire to become qualified in Australia), was concerned with defining properties, surveying flood lines and so on; work in which I was well experienced.

After a month of driving out in the Land Rover, with two assistants, setting up instruments, running lines (we used feet in Australia - a progressive step), I realized that the only real change in my working life lay in the surrounding country. Here it was gum trees, flights of weird birds that cackled incessantly, thorn bushes and miles of brown tussock. We used "slashers" to cut down the thorn bushes; an instrument with a 4 ft. handle and short blade at the end.

One mandatory assignment in that first month was to attend a first-aid course, most of which had to do with treatments for snake-bite. There was no doubt that the Australian bush was a harsh place, with its several species of deadly snakes, to say nothing of very unfriendly spiders and scorpions. Whereas in Canada most snakes would be sporting enough to rattle before striking, the Aussie ones gave no warning, and if - on looking through the transit telescope - you saw an assistant doing a dervish-like dance while pounding the ground repeatedly with his "slasher", you knew that another poisonous one (they were all poisonous) had got its comeuppance.

On Saturday afternoons, I would visit real-estate offices in Cooma, in the hope of finding a house to rent. One day, as I stood on the street, a small Austin car pulled in to the curb. In it were our friends from Sidney, the whole family having driven up from Sidney to see "the Snowy". What followed throws an interesting sidelight on the habits of the Australian family man.

After mutual expressions of surprise and pleasure, George suggested that we go and have a beer (the pub was right opposite). "We" did not apparently include his wife. We spent half-an-hour, downing one beer, and then another. Then, suitably refreshed, we sought the car, where the wife, with good humour, was refereeing the antics of two teenagers and a nine year old boy. Somehow, I was persuaded into the car as well; however, we did not go very far - simply to the next pub (half a block). Here, George suggested that we go and "have another". Which we did; and after half-an-hour emerged to find the wife still keeping the family amused (although not quite so good-humouredly). In the end, after we had repeated the process once more, I decided that I had better take my leave - there were unmistakable signs that the wife was about to explode.

About this time, a furnished house became available for rent in Cooma. The rent was high, but it was Hobson's choice, so I took it, and spent a week-end bringing the family up from Melbourne in the van.

We settled in to our rented house, and next day we registered the kids for school in Cooma North. By now I was travelling regularly out from Cooma to other parts of the project - Jindabyne, Eucumbene and beyond, and was often away from home for a whole week. Kitty dealt with most of the problems on the home front, which included school. This was not Vancouver, where a youngster's introduction to school life consisted of being welcomed by a smiling, young (and usually attractive) grade 1 teacher, under whose tutelage he would have a good chance of enjoying school. The Australian philosophy seemed to be that kids were an unruly bunch, and had to be slapped down regularly to keep them in order. It was reminiscent of my own childhood, with its petty penalties (which are hardly petty to the children themselves). So Grade 1 in Australia was not a happy experience for our son, who took an instant dislike to his teacher. I know it was a great relief to both our kids when they saw the last of that school.

AUSTRALIA 2.

"The older I grow, the more I distrust the familiar doctrine that age brings wisdom".

– H.L.Mencken (1880–1956).

Accents are a subject which I always found interesting, especially in Britain, where at one time it was possible for an expert to pinpoint the very hamlet where a speaker passed his early days, even when such were long gone. On the other hand, upper-class English, the so-called Oxford accent, has nothing unique about it, and can be quite dull to listen to (although it can often arouse a certain amount of suspicion about the speaker himself!)

I found the Australian twang quite agreeable to listen to, but sometimes hard to understand. In fact, with one of my assistants there was a distinct language barrier, and he managed to baffle me two times out of three. Fortunately Nick, my other helper, whose speech was a few degrees less excruciating, was able to understand both him and myself, and act as interpreter.

Nick was a Ukrainian Australian, and a very likeable character. When duck-hunting season became open, he persuaded me to go out with him to one of the local lakes. He had planned it all very nicely. There were no other hunters at this lake (which was just as well). There were a couple of old blinds that we could use, and he even had a few battered-looking decoys (liberally spattered with pellets). He also had two ancient, and even more battered-looking shot-guns. We were to get settled in the blinds in daylight, and when the light faded, and we could hear the sound of the ducks flying in, then we would let them have it. Thus we proposed . . . . . .

In the event, all went according to plan except the final landing. The ducks were a lot smarter than Nick had given them credit for; they waited not till dusk, but till it was pitch-dark before coming in to land. Moreover they came in low over the horizon. That there actually were ducks, there was no doubt - we could hear the swishing of wings, and the landings on the water. The latter, due to phosphorescence, was visible to a small extent, and most of Nick's (unsportsmanlike) shooting was done at water level. On one occasion I actually saw a duck - it was silhouetted against the lights of a car, some distance away on the highway - so I quickly aimed and pulled the trigger for the right barrel. The gun obligingly fired both barrels at the same time. I soon found out that whichever trigger I pulled, the gun still fired both barrels. But the birds were pretty safe as far as my shooting was concerned, as I had an aversion to firing unless I had a target to shoot at. Nick had no such scruples, however, so I figured it would be safer to assume the prone position - which I did, while Nick blazed away, lighting up the whole countryside with his gun flashes. After five minutes of this barrage, the birds stopped coming in (who could blame them?), and Nick and I spent the next half-hour staggering around in the water up to our waists (we had no hip-waders), collecting dead and dying birds.

One of the other surveyors, on hearing of our exploits, told me that he also had accompanied Nick on a duck-hunt - for the first and only occasion - and thought it was only a matter of time before one, or both, of Nick's "fowling pieces" backfired and blew someone's head off.

Socially we did not make connection with many Australians. However, there were several ex-Vancouverites with whom we made friends, and who helped us a lot - notably Fenton S. and George A. Eventually, by the process of "bugging" the executive whose job it was to allocate Authority housing, we were given a house in an area commensurate with my "rank" as an engineering executive. Authority housing was allotted as in Canberra, where subdivisions were segregated for the use of bureaucrats according to their equivalent status. Thus I found myself close to George A., but quite a lot below Fenton S., who was one of the top engineers, and lived in a superior house with a view.

This should have been a joyous occasion for us, had I been able to stay home for a few evenings, to do the necessary fixing after a move. Unfortunately, the Authority felt it was more important for me to be in Jindabyne during the week.

Jindabyne was a village situate on the banks of the Snowy River. In a few years the new lake level would engulf it, so a new townsite was being built further up in the nearby hills. The new layout included public buildings, blocks of residential lots and a shopping centre, and while not urgently required, was - I was assured - of paramount importance.

In all I spent three weeks laying out parts of the Jindabyne Townsite. But how necessary the work was is a moot point, as I found during the second week that all the corner posts whose position I had laboriously computed beforehand, were already in the ground, having been set previously. Moreover, after a phone call to my immediate boss, it seemed that it would have been more politic to have kept quiet about the matter. At this point disillusionment reared its ugly head, as it was obvious that I was not being paid to do necessary work, but simply to "put in time"

It was quite a coincidence that, one week later, I received a letter from David U. in Vancouver. This letter was like a breath of fresh air - Canadian air. In effect it said that the recession was over, and there was lots of work in British Columbia, and that if I were to get my ass up to Vancouver (at my own expense) by the first week in June, there would be a job for me surveying micro-wave sites, in the Monashees along the line of the North Thompson River. For my own part, there was absolutely no hesitation; the decision was made in a matter of minutes - there remained only the difficult questions of *how*. Yet, if we had got to Australia the way we did, we must surely be able to get back.

Obstacles are made to be surmounted. We held the Australian equivalent of a "garage sale", and sold most of newly acquired furniture. To our children's chagrin we gave away our two cats, promising to get replacements in Vancouver. I triumphantly gave in my notice to my boss, and even more triumphantly notified the property manager that we would be moving out of the Authority's house in two weeks. We bought one-way tickets via Quantas Airlines to Los Angeles, with connections via United to Vancouver. We consigned most of our baggage by P & O freight, and kept what we thought was allowable as baggage on the plane; and when we arrived at the Cooma airport, where the D.C.3. was to fly us to Sidney, we were at least 100 lbs overweight. We thanked the Lord for the good friend we had in Fenton.

We shared the plane flight to L.A. with David Sleighton, an astronaut, who gave his autograph to the kids. In two days we were marvelling at the sight of the Fraser River, and all that *water*; and to welcome us home, several old friends were at the airport.

6. HOME AGAIN

BACK TO THE BUSH.

"'Tut, Tut, child', said the Duchess, 'everything's got a moral, if only you can find it.'"

– Lewis Carroll (1832–1898).

The northerly face of Canoe Mountain, overlooking the Rocky Mountain Trench, is a 3000 ft. precipice. Don's method of leaving that summit was to raise the "chopper" up from the ground about five feet, then fly with a smile straight over the edge; and while the machine would proceed sedately and horizontally ahead, my stomach would plunge abruptly (and vertically) all the way to the bottom of the mountain. I never did get used to that take-off.

We were locating sites for micro-wave relay stations, recording lines of sight, elevations, crest clearances and topographic data. Including myself, we had a crew of three, two long Land Rovers packed with gear, and a helicopter with pilot and mechanic. The two men on the crew, Tim and Ralph, were both keen and likeable guys (no complete disinterest, or chips-on-the-shoulder here, as in my last billet), the country was exhilarating and the project challenging.

1962 was a very mixed up year. In May I was surveying in Jindabyne Townsite in New South Wales, in July I was on top of Canoe Mountain in Central British Columbia, surveying for micro-wave sites. In the spring our children spoke "stryne", in the summer they had to learn

Canadianese all over again. But it was in a good cause - the nomadic existence was soon to be over.

We had friends - how many we did not realize until we came back - and with their help and my wife's dedication we sorted out our family matters in short order. In a week we had a car, and in ten days a furnished, rented house on Cornish Street. The idea that houses could actually be rented for a reasonable cost was a notion that we found strange at first, after the past year of frustration. How could we not have realized how fortunate we had been previously.

So, while Kitty picked up the threads of homelife again, I went off once more to work - I had hardly stopped doing so since our ill-starred experiment began. But at least one could now stop worrying about the future. In fact, the present was going to take up all my time for the next three months.

This job was in complete contrast to the experience of the past six months. Although the Snowy Mountains of New South Wales were considered to be "outback", and the terrain was somewhat harsh, working conditions were positively suburban compared to those in the wilderness of British Columbia. This was real wilderness, rugged, wild and steep. I have always found the mountains of B.C. intensely beautiful, but whereas most people are content with merely looking at them through the window of a car, I find it most satisfying to be in amongst them, and in this case I was being paid for it.

The North Thompson highway of those days went the same route approximately as Highway #5 does today, following the valley by way of Clearwater, Vavenby, Avola, Blue River and ended near Valemount. But although the general direction was the same, this was the only similarity between the two roads. That old road was a bone-shaker, and in places it resembled a creek-bed. At Valemount it joined another, even rougher, road to Prince George, and a reasonably good road to Jasper. It was on this North Thompson road that we broke a main rear spring, when going to the Rockies with the family in 1960. Now it was 1962, and the road was if anything slightly worse.

The network of proposed micro-wave sites followed this valley, with inter-visible stations on high mountain ridges or peaks, at intervals of several miles. The hotels at Clearwater, Blue River and Valemount saw us in the evenings; I do not remember any hotel at Avola. These hotels were not modern (there was no need, as the road attracted few tourists).

We spent the days, and even some evenings in the hills; we back-packed, cut out landing places and did all the necessary activities associated with survey work in such places. When the lighting was poor, we used the "chopper" as a distant signal, hovering vertically over marks that otherwise would be indistinct. We used it for dropping in power-saws, equipment and lunches. Never had I worked with such a useful device. When Don, the pilot, was not being used actively, he would fly down to a neighboring ridge, stretch out and work on his sun-tan until quitting time; at which time, he would ferry us in turn down the mountain, where we would very soon be sitting in front of a (well-earned) beer. Work in such places I had done many times before, but always there had been, at the end of the day, that five or six mile bushwhack to get back to camp. And usually camp would be literally a camp, located for convenience to the workplace, not for any notions of comfort.

This same northerly face of Canoe Mountain, when we first arrived, sported a spectacular cornice, and curiously, a surveyors stake, which seemed to have been planted about half-way beyond the edge of terra-firma. I often wondered who, in blissful ignorance, had set it. Needless to say, we did not disturb it. The southerly slopes of this mountain were of a good gradient, and would have made a good ski-run. We were required, towards the end of the job, to locate a jeep road from the valley to the top. For this we did not need the helicopter, and we waved him regretfully away one day in August.

At the time we were staying in Valemount, and driving out each day to the foot of the mountain. At the end of the first day, as we drove back, we passed a small cabin just off the road, and by a creek. The idea of a cool beer at the hotel had been in our minds, but the lady who was at that moment watering some plants in front of the cabin forestalled us. "Come on in, boys, and have a beer", she said. We were immediately persuaded.

Her name was Gail, a not unattractive and definitely independent gal. This was her place, which we admired; and where she lived on her own when not working, did a bit of trapping, and fishing, and grew a neat little garden. If there was a boy-friend we never saw him. At other times, she worked as a cook on the railroads, and in bush camps.

It was on the way back to the hotel, and the usual unappetizing meal, that inspiration came. Next morning, as we went by the cabin, we made a deal with Gail; that if we bought the grub, she would cook it for us, for a fixed rate of so much per meal per person. I will not even quote the exact amount, because it would seem completely ridiculous at today's prices; however, she thought it was a great idea. Armed with her grocery list for the week, we bought groceries, and delivered them to her. After which there were absolutely no complaints about food. We were there for breakfasts and dinners, and she made up huge lunches. I remember Tim eating five eggs at breakfast, and half a pie for dinner; and there was always more; while she regaled us with stories about life on the C.N.Railway, as seen from the galley chef's point-of-view. Her language included most - but not all - of the four letter words. Tim gazed at her with open mouth, never having met anyone quite like this before. She called me "Sweety!" at times, (to Tim and Ralph's delight), and we responded to her kidding by eating everything in sight.

We certainly needed the calories whilst slogging our way up Canoe Mountain; which - it became apparent - we were sharing with another animal. The fresh bear signs we kept finding were big enough for each to fill a laundry tub, so it must have been a very large grizzly; but by dint of making as much noise ourselves as he did, while he crashed around in the bush, we each managed to avoid each other.

Somewhere on the road to Clearwater one of the Land Rovers broke its tie-rod, leaving both front wheels pointing in opposite directions. My memory is hazy as to how we dealt with this problem, but somehow we towed it, left it at a garage, caught the train to Vancouver, where after a few days off, we acquired another vehicle to drive back. This was now the end of the job, and in Vancouver a week's office work awaited.

I found my wife in the rented house, and my kids in the local swimming pool, already jumping off the top diving board. Considering that when I had last seen them in the water, in Australia, they had only just learned to swim, this was quite an achievement. In fact, this summer had marked an achievement for us all.

TWO GREMLINS.

*"Things which you do not hope happen more
frequently than things which you do hope."*

– Titus Maccius Plautus (254–184 BC)

A gremlin is a minute person - perhaps like a leprechaun - although not necessarily colored green; but whereas a leprechaun can often bring good fortune, a gremlin usually brings nothing but plain dirt. We were to find some in North Vancouver quite soon.

In Vancouver, it was back-to school time. We enrolled our kids in a brand new school, where not just our son and daughter were new but all the students were. We rented another furnished house nearby. Having got thus far with our arrangements, we were taken unawares when the first gremlin stepped nimbly from behind a bush, and (to mix metaphors), tossed a spanner into our nicely ordered works.

It seemed that although Canada had started to recover from its 1960/1961 recession, the recovery was in the nature of a false start. The real recovery would be another six months down the road. Had I been able to scrutinize the Financial Post during the past summer, rather than the mountain tops, I would have known about this unexpected setback; it meant that employment for surveyors in the winter might not necessarily be a foregone certainty. There followed a period of intense activity on my part, during which I relieved another surveyor who had not had a holiday for two years (this sometimes happens in the surveying profession). I finally zeroed in on a partnership in an old-established firm of surveyors in Vancouver City.

This was to prove the best investment I could ever have made The senior partner of the firm, whom I was buying out, was soon to retire, and his son, who now became partners with me, was a man whom I had always found likeable. I say deliberately that Hamish always had the auspicious knack of getting along with people. I seem to have landed on my feet again, gremlins notwithstanding. The fact that 90% of the work was city work - was not a deterrent at all. If shopping centres, apartment blocks and building certificates were to be bread-and-butter, I would look forward to it all, and keep an open mind.

The work of a land-surveyor had been changing ever since I had first become qualified in 1950. In the old days, a surveyor had been a pioneer; his business was in the woods, where he explored, surveyed preemptions, Crown Grants, mineral claims and often did comprehensive surveys involving both the engineering and legal aspects of large projects. What little work there was in the cities was jealously guarded by firms such as the one of which I was now a part.

But with the end of the Second World War, every city in British Columbia began to grow at an unprecedented rate; so much so that surveyors were hard pressed to keep pace; and the city and suburban work, which now seemed to stretch into the distant future, became the surveyors livelihood.

This firm, into which I had bought, was not merely an old firm; it had the distinction of being the oldest survey firm in the City of Vancouver, having been established shortly after the first transcontinental train arrived in 1887. It's past history included the original layout on the ground of the downtown area, West End, Hastings Townsite and resurvey of the Old Granville Townsite - the latter project at the request of the Provincial Government after the fire of 1886. The original controlled layout of a city determines for all time how well property lines can be redefined later. Vancouver was most fortunate in having three qualified and conscientious Dominion Land Surveyors, to lay out such precise lines in the last few years of the 19th century. In old files at the office, there were records of all the major engineering projects of those days in which they were involved, much of it now having been given over to the City Archives. Searching through the files, as I did in the next few years, became for me a most interesting extra-curricular activity.

The retiring partner in this firm was a most interesting man, and he remembered a lot about early Vancouver; and also the depression days of the '30s. At that time, when there was so little work that most surveyors quit the profession completely, he managed to keep going, running a small business until the 1960s. We were to keep this city practice going for about two years, before extending our horizons.

There were two government offices which it was essential to visit before undertaking a legal survey in the City, the Land Registry Office and City Hall. Description of the City Hall Engineering Dept. cannot be complete without mention of Hecky, who was city surveyor for many years in the '60s and '70s.

Hecky was a stalwart and most likeable man, who was always ready to help anyone who came to see him. He presided over a roomful of draftsmen and assistant surveyors, and was seldom seen at his desk, preferring to wander up and down and talk loudly. His talk was unusual in that most nouns (and often verbs and adjectives) were well-embellished with expletives. One soon became used to this; it was just part of Hecky's personality.

Whether someone in the Personnel Department wanted to embarrass him, I'll never know (in any case it would have taken more than it did); but one day City Hall hired a new draftsperson, and it was a woman, who duly took her place in the office. A few days later, while Hecky was parading up and down and sounding-off, someone suggested to him that since he now had a woman in the office, maybe he would have to curtail his language.

"Language - what language?" replied Hecky, "Hell's bells! - that's my -----------ing draftsman!".

After Jim retired, we moved to a bigger office, and enlarged the scope of our activities. But there was a second gremlin lurking in the shadows.

We had hired an efficient secretary, and a full-time draftsman, yet work continued to increase; and in the end - I started taking work home. Overwork was just what that mean little dwarf was waiting for.

Hospitals, in my opinion, are places to be avoided, particularly if one is a patient. When walking around, visitors can easily gain a misleading impression, since the beds look so neat and tidy, the nurses so well-starched (and sometimes well-stacked), and the patients so well-relaxed (with nothing to do but watch television and have their meals brought to them). But as a patient myself, I did not find any of this meaningful and the only good experience which I remembered was the news of my impending discharge.

During my sojourn in bed, there was one small incident worthy of note, which illustrates how astonishing can be the results of the unintentional mixing of two otherwise predictable drugs - "chloral hydrate", the old-fashioned sleeping-pill, with another drug prescribed as a pain-killer. The "high" that resulted was nothing short of miraculous, and since miracles don't happen so often these days, I think it should be recorded.

Before going to sleep, I was breathing with difficulty and with some pain - (I do not recommend pleurisy as an illness to be contracted if one simply needs a rest). During the night I had a singular dream, in glorious technicolor. I seemed to be perched on the edge of the grill of a huge, black barbecue, which held bright red coals, and the lid of the barbecue was slowly, slowly closing. So obviously it was going to be necessary for me to move fairly soon. When the proper moment came, I did move; in this case, right over the end of the bed, where (this also seemed part of a pre-ordained plan) I started sprinting, bare-footed and bare-assed, around the wards, and between beds full of sleeping patients. Eventually I was neatly fielded by two night-nurses, who firmly led me back to bed, murmuring outraged admonishments. ("You really mustn't run around like this; you are a naughty boy"). All this I remember subconsciously, but it seemed in my awareness, that it was happening to someone else, while I was merely the observer. The fact remains that, had I not been under the influence of such a mixture of drugs, I could not even have sat up in bed unaided, let alone done an imitation of Harry Jerome dressed in a short hospital gown!

There are other recollections of that session in hospital, some pleasant as when visitors came, especially my wife, who was ever present; and was instrumental in getting a "second opinion". And some unpleasant, as when a practical nurse, who knew nothing of my condition, determined to make my bed. I could hardly blame her for carrying out her instructions; on the other hand it would be necessary to roll me over in order to do so. I tried to explain why I could not be moved; but what does one do, when the obviously uninformed lady takes no notice; something had to be done, even if it consisted of summoning all the breath that I could muster, and requesting her in words of four letters to "* * * * get lost!". (It sounded like the croaking of an asthmatic frog). Fortunately, my roommate then pressed the bell, and another nurse came to my assistance. Next day found me in an oxygen tent, and from then on conditions (including mine) started to improve.

Six weeks I had been out of action. But I was still around. I believe Kitty had thought, at one time, that I might not make it.

JUST EARNING A LIVING

"You cannot teach a crab to walk straight"

– Aristophanes (450–385 BC).

One of the hang-ups inherent in the make-up of surveyors who were nurtured just after W.W.II, was that they never turned down work if they could possibly help it. In the expanding economy of the 1960s and 1970s, this led to a lot of "busy-ness", and caused us to add a third partner to our firm - my ex-pupil from former years, Alec. Now, we naively thought, we will not have to overwork ourselves any more.

I find writing about this period more difficult than earlier days. Because, in spite of the variety and scope of the work that saw our transits eg:- condominiums, shopping centres, swimming-pool complexes, office buildings, subdivisions, rights-of-way for highways, transmission lines, ski-lifts (you name them, we did them) - we had become so established that we would look upon a job as more of a business proposition, and less as a venture. Yet surveys done in the "real" old days always have given me a sense of being venturesome.

City and suburban surveys, however, become in the end routine, and are memorable on only a few occasions - such as the time a drunk walked right through my transit, which was set up at Hastings and Abbot Street, then disappeared in the traffic before even any sound had time to emanate from my outraged, open mouth. The transit, well and truly bent, was a sad sight, but fortunately insured. Now *that* was a memorable occasion, even if it was one I'd just as soon forget.

Still, there were other perils to be avoided besides bent transits when doing work in the city, and this one should be mentioned if only to show that the age of chivalry ain't what it used to be.

It happened in the West End of the City of Vancouver, where in the sixties high-rises were sprouting like mushrooms, generating a lot of work for the local surveyor. On this occasion my assistant and I were going about our normal business, when a lady crossed the road in front of us, and then fell flat on the sidewalk. Harry my helper, was fairly close to her, and I was quite puzzled to see him pointedly ignore the incident - however, it seemed to me that the poor old soul might be hurt, so I approached and said, "Are you OK? Maybe I can help you . . . .", then made the fatal mistake of putting my hand out.

Too late - too late! The "poor old soul" suddenly came to life, encircling my wrist with a bony hand and vicelike grip, and directing a breath at me which almost felled me to the pavement.

"Thass alri', honey", she crooned, "I jus' live 'long here".

So thither I had to escort her, and waited while she fumbled for the latch-key with her other hand (once having snared a man she wasn't about to let go with the first one), and finally opened the door.

"C'mon in, honey", she murmured with a hiccup.

"Thank you, but no' - I replied, as I wrestled my arm away, and closed the door. Behind me was Harry, with a smirk from ear to ear on his silly face. But he already knew what I had just learned. Damsels in distress should be left well alone.

Now it was a new era, and since we were taking on sizeable jobs, it was only natural that we should be one of the first to try out the new equipment. Electronic measuring devices (EDMs) were just coming into use at the time. Although unwieldy at first, they were revolutionizing the technique of field surveying. The first ones weighed 50 lbs, while the tripod, transit and battery weighed another 50 lbs, plus axes, reflectors, range-poles (and lunches). Obviously one could not be too mobile except near a road.

It was on a road that we had our first experience of the new equipment. The old tried and true method of measuring centre-line control was by means of a 300 ft steel tape. We were always conscious of drawbacks to this system, not the least of which was the tendency for a certain type of motorist to look upon the sign "Caution - Survey Crew" as a signal to accelerate. Presumably he hoped that - if he were lively enough - he might prang the instrument-man before he could leap to safety; or at least run over the steel tape before it could be pulled out of the way. While such a method promoted a variegated vocabulary on the part of the crew members, it was also quite time-consuming, especially as the line always had to be measured twice, as a precaution against errors.

So for our next highway centre-line survey, we hired an EDM, and its operator from another firm. The operator turned out to be Tim, who as well as demonstrating its possibilities, described it all as a "piece of cake". We were quite easily sold on the idea.

In view of the cost of this equipment, it was incumbent to have "in the bag" a fairly extensive project, the proceeds from which would help pay for it. It seemed that the economy of B.C. was working well for us during those years, as the Right-of-Way surveys for the Peace River Transmission Lines came right down to the Lower Mainland, and a good many survey crews were required for it.

I must say that to have undertaken some of those particular jobs *without* benefit of an EDM would have been as tough as it was impractical; and I imagine we were not the only firm who acquired an EDM that year, and thus entered the Space Age.

Naturally, there were drawbacks. The bureaucrats, who abhor a state-of-affairs where problems can be easily dealt with, decided that we needed a new headache (to replace all the old ones). They introduced the metric system. Although the switch-over is easily handled by any calculator, the fact remains that all existing official plans at that time were dimensioned in feet and decimals, while many older ones were given in chains and links; the latter now requiring two transpositions; much of it needing to be done by the surveyor when in the field. Happily, even that chore has now been simplified, as most modern theodolites now incorporate transit, EDM., and computer all in one instrument (a "total station"), which has appropriate buttons to push, and is easily portable by one crewman.

It is ironic to think of the hours that we slaved away, high on the mountainside in those long-past, adventuresome days, with transit, steel tape and a book of logarithmic tables. All that is required today is a pocket-sized "total station" and a sturdy, well-educated index finger.

THE HUNTERS.

"I hate quotations."

– Ralph Waldo Emerson (1803–1882)

Ray was one of our most interesting neighbors when we lived in North Vancouver. Certainly he was the most convivial; that side of his nature being stimulated to a large degree by his association with another interesting neighbor - Manuel.

Manuel hailed from Spain, and came from a family which was well acquainted with the grape. In his basement, adjacent to Ray's house, he had numerous barrels, containing wine in all stages of evolution. Every year in the fall he held a wine-making party, to which the neighbors were invited. He used a rented crusher to squash up the many kilograms of grapes, which he had imported from California. The neighbors were there, not particularly for the purpose of giving assistance, although we might occasionally take a turn on the handle, but mainly to celebrate the festival of the grape. Wine from various barrels would be sampled and commented on, and a good time had by all. Manuel's wine, though not strong, induced a real sense of relaxation when sipped for a couple of hours. I always knew when I'd had enough because my eyes would involuntarily start to close, and if Ray, who was an inveterate talker, started having difficulty with his words, I knew it was time to go home.

It was at one of these happy ceremonies that Ray inquired of Manuel what happened to the "mash" after the wine had been siphoned off from the primary fermenter. Manuel was wont to throw it into the garbage can, or on the compost heap; however, he was quite happy to give it to Ray; and Ray, being an old Yukoner, put it to very good use.

The liquor that he produced with the aid of copper coils and other apparatus in the corner of the basement, was basically "grappa", a first stage in the distillation of brandy. It was colorless, burnt with a blue flame, and when a person drank it, it went down with a slow burn (followed by a hiccup). Ray was very proud of his moonshine, and offered it to all and sundry visitors to his house.

What all this has to do with hunting will become apparent in due course.

The road to Whistler had not long been surfaced, and at that time there were quite a few properties for sale in Pemberton and beyond, in places where, a few years before, the B.C. Railway had been the only access. Gravel roads led from Pemberton, through Mount Currie, to Devine, D'Arcy and Anderson Lake; interesting places, all ripe for summer cottage development.

Ray itched to acquire a property where he could get away at weekends. Before long he had bought a Crown lease on Birken Lake. As a wheeler-dealer he had no equal, and by the time I was invited up there, he had bought an old bunkhouse from a logger in Devine, cut it completely in half, had it transported to the site, and set the two halves together on piles by the water's edge. I got the impression that all this work was done for next to no payment in actual money. He ended up with a cottage-like building, perfectly suited for a week-end retreat. He put a dock out into the lake, which was warm enough for swimming in summer, and contained a fair number of rainbow trout (all waiting with open mouths to snap up any lure which Ray might throw them).

On the several occasions I visited there, he introduced me to various neighbors. One of them, Clem, kept several horses on his ranch, one of which he had sold to Ray, but continued to look after, so that Ray could go riding when up there. Ray's horse was a very large horse, at least sixteen hands, and well-built for mountain trails. I don't know if it was Clem's suggestion that Ray get a hunting party together, but in any event, one late summer evening found Ray making enthusiastic plans for the Labor Day week-end.

I have always felt that for a would-be hunter to leave town on a holiday week-end, go out into the bush, successfully kill a deer or a moose, and return on the third day with the meat, is a very ambitious project. Although I fancied myself as red-blooded as any home-grown Canadian boy, I had never been hunting in British Columbia. I had met many that had, and could pretty well divide them into two groups - the hunters, and the would-be hunters. I figured that Ray himself was probably one of the former. So in preparation, we both spent an afternoon at one of the Burnaby rifle-ranges, zeroing in our rifles; mine was an ex-army .303 calibre, adapted for sporting use - Ray's was a much better gun. But I figured that the other two members of the party, Manuel and his friend Roberto, were like myself of the "would-be" variety, and I had some doubts as to how bush-wise they might turn out to be.

However, we had an excellent guide - so not to worry. Clem had cut a trail for the horses up to 7000 feet elevation, above Anderson Lake; once up there we would be able to take our choice of deer - in one direction, or goat - in the other. Both (like the fish in Birken Lake) eagerly awaited our arrival.

We drove up Friday night to Ray's cabin. Now I admit that often my ideas are somewhat far-out, but it seemed to me that going hunting would usually involve starting early in the morning, probably at first light, and, to that end making sure that everything was in readiness. In these preliminaries, however, it seemed that I was far-out. Ray's ideas were quite different, and included inviting everybody in the neighborhood to the cabin that night, for a wing-ding of a party (to celebrate going hunting). The booze was free, as it was largely moonshine, although Manuel and Roberto and myself all brought rum as well. We were *very* well-equipped with alcohol. However, the guide, who was unused to moonshine, was as sick as a dog during the evening, and it was obvious that he would have problems getting up in the morning.

In fact, it was twelve noon when we finally dragged ourselves out to Clem's ranch, only to find that he was still recumbent. However, he was a good sport, and in spite of his hangover, he organized the horses, and helped by his wife packed four of them with food, sleeping bags and assorted goodies.

Although all the other horses seemed to have names, mine was referred to as simply "that roan horse". However, Roan Horse and I got along very well indeed. He was a very rare animal, in that he seemed to understand all my signals, and was not slow to carry them out. What more can one ask. Manuel's horse was an Appaloosa, bridled with a hackamore, a going concern of a horse; he took the lead without question, and I figured that if the slope of the hill had been 90 degrees, he would still have tried to go straight up it. Manuel, incidentally, was a pretty good rider. Roberto was a very tall man, and he was riding a small horse. In fact, Roberto, with straw hat and a cigarette drooping in the corner of his mouth, his feet almost touching the ground, and a rifle dangling from his shoulder, could well have doubled as the Cisco Kid. Ray's horse was, I am sure, part Clydesdale; but when one saw them together, one realized that they complemented each other very well; Ray himself being built much along the lines of a cart-horse, also.

The party started in reverse order. Manuel was in the lead, followed by the rest of the hunters, and the guide was in the rear pulling two pack-horses. The other two pack-horses were in the middle, unroped, and following the horse in front of them. Clem still had a hangover.

It was a steep, narrow trail, and we zig-zagged up the mountain in the heat of the day. About half-way to the top, at one of the steepest and narrowest places, one of the unescorted pack-horses stumbled, then left the trail - arse over tea-kettle down the hill, and by some outlandish stroke of luck, came to rest on his feet up against a tree some forty feet below; distributing his pack boxes and their contents all over the slope on the way down. He was, poor animal, very much shaken up, and I went down to hold him, but he was uninjured and simply stood there, looking sad, as well he might have felt. As for Ray, he was even more shaken up, as the boxes had contained two half-gallon jugs of grappa, now completely smashed. I remember thinking that as far as that consideration was concerned, the accident was probably all to the good. But in fact it didn't matter at all. There was lots more of the stuff in the other pack-boxes.

The trail was actually too narrow for a horse to turn around, although Ray tried, and we were treated to the sight of his horse, sitting on its hind-quarters, with its back to the hill, rather like a dog, while Ray slowly slid backwards out of the saddle.

At the top of the hill we made camp, had some food and wine (naturally we had wine, too) and made ready for the night. I remember Ray complaining about the broken bottles, and Manuel saying: "Well, a good thing it was only grappa, and not the wine. Because, if there's no wine., I don't go hunting, I go home!"

We awoke to the murk of a pea-soup fog. Perhaps it would be more accurate to call it a thick mist, pea-soupers being reminiscent of Manchester, England in December. But it was just as thick as the Manchester variety. Visibility was limited to fifty feet at the most.

We ate breakfast; after which Manuel and Roberto elected to go with Clem to a high vantage point, where they planned to wait for fine weather (and goats) to materialize. Ray and I, after a short trek to another hilltop, and an hours vigil, during which the cold penetrated as far as our bones, decided to return to the camp-site and make some improvements to it; thereby keeping ourselves warm, and providing some comfort for the others when they should return, frozen to the eye-balls.

With a small axe, we cut poles, and made a 25 ft. long lean-to framework, and filled in all the gaps with spruce and fir boughs. It was, although I say it myself, an excellent example of woodmanship - a perfect shelter, with the fire beaming in the heat from in front of it. It would have been even better if the wind hadn't changed as soon as we finished it. However, the rest of the party - when they arrived at 3 pm with teeth noticeably chattering - didn't complain at all.

We spent the rest of the day in our sleeping bags, only getting out to make supper, and for nature calls, and to pass round the wine, or the rum, or the moonshine. The stories, also passed around, progressed from being good-humored, to hilarious, to outrageous, then to complete fantasy, and finally to drowsy incomprehensibility. Again we slept.

We woke to bright sunshine all around. A day to elevate the soul's awareness, had it not been for that little man with the hammer, who had taken up residence inside our heads. As we wolfed down the pancakes which the resourceful Ray had cooked up, we could look out on the mountain-top view, which we had not seen properly until now, and wish that we had another day up here; when we could *really* go hunting. But that was not to be; we had to pack up, get ourselves down the hill and back to Vancouver. All of us were due at work the next day.

I have to admit that we *enjoyed* the trip. What did we have to lose? Nothing. What did we gain, apart from saddle sores, sun burnt lips, insect bites and hangovers? That is a hard question, but there must be an answer; otherwise why would so many men continue to go hunting year after year?

CHANGE.

"The more things change, the more they remain the same."

— *Alphonse Karr (1808–1890)*

I have heard people who have nostalgic memories of a place say: "Never go back!". In this case, however, since my memories are not wholly nostalgic, it might be interesting to compare 1953 with 1972, when I made another working trip to Masset, in the Queen Charlotte Islands, and to other coastal places.

This time there was no S.S. Cardina to wallow its way across the Queen Charlotte Strait. The Union boats were long gone, the only reminder of them being the legacy of derelict government wharfs, whose drunken pilings, distributed throughout the inner coast, now served no useful purpose. I found the most significant change to the B.C. coast in 20 years was the spectacle of deserted docks, abandoned houses, gravel roads overgrown with alder and willow, garden fences covered with salal, and the once neat gardens themselves full of nettles. The early 1950's, when I had worked here, had been the tail end of a prosperous era of the coast, when there were scattered settlements in most inlets - scattered maybe, but they were *settlements*; and when the Union boats were in their heyday, every log-dump in the inlets could rely on a service of at least once every two weeks. But progress cannot be denied, and the steamships finally made their last runs. Now we fly everywhere, and the freight goes by barge; and while the settlements withered away gradually (leaving behind almost unbelievable tales of independence and adversity - although the settlers themselves would probably not have described their lives in such terms), the towns grew into cities; and while the same logging and fishing, and connected industries still thrive, they are controlled from coastal towns, where people live in subdivisions, fly out to work for five days, then return on week-ends to mow the lawns. Of course the number of boats has increased enormously, but much of the increase is due to pleasure craft, and there are fewer camp tenders for the simple reason that there are fewer camps.

These were the biggest changes I found in the early 1970's. But Masset's change was quite predictable. It now had a modern hotel, motel, restaurant, liquor store and public library; moreover the cows had disappeared (that was bound to happen). It was hard to associate this modern town with the village of 20 years before, with its pot-holed streets and paint-peeled houses; and just out of town was the most praiseworthy project of all - a bird sanctuary, in which swans, cranes and herons strutted, safe from human and other predators.

Some things had not changed, however. The great church with its huge totems, which dominates the older part of the First Nations village, still seemed to belong in a different world. In 1953 I found it to be the same - a timeless almost alien setting, and certainly not one to be affected by our ideas of what constitutes progress. Yet other parts of the village, at that time, with their dilapidated roofs and cluttered yards, the garbage and starving dogs - those could be measured in our terms, and they denoted a particular lack of progress, of which Canadians should be ashamed. No country as well-off as Canada should allow native people to live in such neglected fashion. While New Masset had indeed changed for the better, it was noticeable that such progress had not made its mark on Old Masset.It was quite indicative of the change that, while in 1953 the settlements would be referred to as "New Masset" and "Old Masset", in 1972 such designations had become - "Masset" or "the Indian Reserve".

Another forty years have now passed since 1972, and while the white man's world has continued to develop, that of the native peoples in this Province continues to stagnate.

In the 1970s, we had occasion to resurvey several Indian Reserves in far-away places, and the "Connoisseur" was the answer to our problems - she transported us to the jobs, and we lived aboard her while working. She was an ex-Newfoundland schooner, with a deep keel, fitted with diesels, which had been used as a fish-packer. She was 100 ft. long, had all the navigational aids, and would do about 7 knots. She came with skipper/owner - Grant - his girl-friend as cook, and we chartered her.

I had an aversion to using chartered planes to do small jobs in remote places, because often one little detail can cause the whole plan (and project) to go amuck. For example, when the pilot finds that he cannot take all the baggage; or where one is in such a hurry to take off, that the transit - or the tripod - or a briefcase containing all the plans - gets left behind on the dock. Believe me, *all* these things can happen, and *do*. But the biggest problem occurs when the job is finished, and you are sitting on your bags awaiting pick-up, but the plane doesn't come . . . . and still doesn't come . . . . although the weather is perfect . . . .

We had recently done two surveys from a place called Boughey Bay in Port Harvey. The designation has nothing to do with boogies, at which I was a little disappointed. Actually, the name is quite prosaic, commemorating one Lieut. Boughey, the 1st Lieutenant of H.M.S.Havannah. But although our stay at Boughey Bay's float camp turned out well, chartered planes, and chartered boats included - worry-wart that I am, I was all the time expecting something to go wrong.

The next year would be different; and that was where the Connoisseur came in. In May of that year we loaded survey gear, two week's grub, personal gear including fishing-rods aboard, and set off for that summers project, Port Neville.

One aspect of the B.C.coast had not changed. Mostly we were working in unlogged areas, and the old growth trees were the same. The underbrush in which we wallowed while cutting lines from A to B was just as thick, maybe thicker; and it was just as wet even though it was summer. The tide flats were still as interesting, and there were still oysters and clams, and abalone at the right tides, since we were a long way from the cities and their attendant pollution.

At Port Neville we had another example of change. This one became apparent one night at the flood of the tide, when we let down a gill net, in the hope that we might snare a salmon or two (after all there were two deep-freezes aboard, containing only some tommy-cod and oysters, which don't count as real fish). At 11.30 pm I went to bed, and slept undisturbed by the flailing around of heavy boots on deck. At 7 am I awoke, and went topside, to find a deck knee-deep in dogfish; literally hundreds of them. Most of them had to be cut out of the mesh (it was an *old* net, fortunately). We spent two hours throwing the dog-fish overboard; many had young ones alive inside, and they swam around with gusto attached to their yolk sacs; a lesson in biology. But Buddy, who was a New Zealander, was upset at the waste. "We call these sand-shark, back home ", he said, "and eat them". So we froze half-a-dozen of them, and tried eating a couple of fillets that night. Even Buddy agreed that they were horrible.

Now, no self-respecting British Columbian will eat a dog-fish, and as I found out later, there was good reason. The dog-fish, like the shark, urinates through the skin, and if the skin is not discarded immediately after being caught, the fish tastes absolutely loathsome. So the answer to Buddy's dilemma was that we should have got rid of the skin; when we did they tasted reasonably good. As I pointed out to Grant, dogfish are used in Britain for "fish and chips". (Grant's reply was: "What do you expect? They'll eat anything over there!")

The second voyage of the Connoisseur was memorable because it took us to a place that has seen few visitors. The following year, after a rendezvous at Minstrel Island, and two small surveys, we made for Slingsby Channel. This was the threshold of our journey, as it was close to the Nakwakto rapids, a tide-race that runs for half-a-mile at about 23 knots, with slack water twice a day for no more than 30 minutes. As our worthy "Connoisseur" could muster at most 7 knots, we anchored and watched the maelstrom go past, till it calmed down enough for us to proceed. This channel is the only entrance to hundreds of kilometers of inland waters - Seymour Inlet, Belize Inlet, Alison Sound, Nugent Sound. It was a lonely cruise; sheer cliffs and spectacular waterfalls lined the shore; there were no habitations, no wharfs and no other ships. There were some signs of former human activity - hillsides covered with deciduous growth, having been logged 20 or 30 years before, and other signs - the odd landing stage, rotten, or with the remains of a tank lying beside it, an old A-frame on the shore, or a derelict donkey-engine amongst the second growth. This was Belize Inlet.

I do not know the reason why I am drawn to wilderness areas, I only know that I find satisfaction, and a sense of mental balance when I am in natural surroundings which have never been subject to modern man's dominance. Such a place, both forbidding and impressive, was the head of Alison Sound. Here we were a long way from anywhere, hemmed in by miles of unforgiving, wild country. The silence was vast, broken only by the sound of a distant waterfall. It was a perfect estuary. The water in the inlet was fresh, without a trace of salinity. The river flowing into the bay was sixty feet wide, twelve deep, with a gravel bottom and crystal clear. One could have seen a fish, had there been one, but it is my guess that no fish had been up here to spawn for many a year.

That was the anomaly about the area that disturbed me. There were perfect spawning grounds up the river, but no sign of fish having spawned; no sign of other wildlife either, no bear, and not even beaver. But the fact remained that once many years ago, the head of this inlet was set aside as Indian Reserve, and the reason that native people lived here was because of the salmon. So - the salmon were gone, since we found them so easy to catch at the rapids - and although man had not been active here for 20 years, they had not returned, nor had the other wildlife that depended on the cycle.

A dismal reflection? Give us another 30 years, and the salmon will have gone the way of the dodo and the passenger pigeon; and we will just have to get used to the idea of eating dog-fish.

FAMILY OUTING.

"What the meaning of human life may be, I don't know; I incline to suspect that it has none."

— *H.L.Mencken (1880–1956).*

Jim Flickel's horses were a nondescript lot.

Rastus, who I am sure operated on one lung, had a tendency to look suspiciously at every stump alongside the trail, and I fully expected him to jump sideways in passing it. My daughter Wendy's horse - Star - big, white, strong and very good-tempered, had super-sensitive ears; he had no bridle, simply a halter. Wendy was happy to sit on his back and hold the rope. Foxy was ingenious, as his name implied, and he was just the right size for David, my son. The best of the bunch was Gypsy, the pack-horse, although when David, at 10 years old, took his turn at leading him on foot, it was often uncertain as to who was leading who.

The idea (mine) that we should have three saddle horses for four persons was not for reasons of economy, but because people who have not ridden for some time get quite saddle-sore after 18 miles; and we (that is, I) figured that with only three riding horses, each of us would perforce have to take a turn on foot, thus giving his or her posterior a rest; and logically, whilst on foot, the same person could lead the pack-horse. Theoretically a good plan, but in practice it resulted in myself riding, *and* pulling the pack-horse for most of the way. This I had hoped to avoid.

It was our first real holiday in the Bridge River valley. All previous visits to this paradisiacal world had been in the form of long-week-ends, and we hardly had time to unship the life-jackets before it was time to go home. This time we had two weeks, and we resolved to check out the rumors of this wonderful lake, 18 miles away by horse-trail, where one could camp at 5500 feet elev., amid alpine meadows, and where one could catch all the fish one could eat in half-an-hour. I had visited many lakes in B.C. where similar claims were made, so I was not about to believe this tale either; but it did sound like a good place for a family outing. We made sure that we had plenty of bacon and other staples, *just in case* we didn't catch too many fish. I also had a good map, although Jim Flickel didn't seem to think it necessary.

"Just follow the trail, the horses know the way!", he said, fortunately not adding, as they do in Britain: "Yer cahn't miss it!" - which is an invitation to get lost. But my family never even considered that contingency. After all, surveyors never get lost. In any case we had camped in the great outdoors many times, even though this was the first time our kids had been on anything like intimate terms with horses.

The Gun Creek trail leads up on the north side of Gun Creek. An admirable old trail, that had once been a mining road, with the boisterous waters of Gun Creek on the left hand side, the lofty slopes of the Pearson and Eldorado watersheds on the right. The first seven miles were delightful; lupines, columbines and the occasional wild tiger-lily lined the trail, the sky was clear, and there was enough breeze to discourage the bugs. The horses liked the idea of all these wild-flowers too, but for a different reason, and it soon became apparent that we would need both a leader *and* a follower, the latter on foot with a substantial willow stick, to keep the group moving. So we progressed to Eldorado Creek, where we had lunch.

The sight of the first slide is quite a shocker to a person seeing it for the first time, although from the semblance of a trail crossing it, the route was fairly obvious. Nevertheless, both horses and riders, of both sexes, behind me made up their minds immediately that nothing would induce them to try it, and nothing I could say about how much safer it was on a horse (with four legs), than on foot (with only two), made any difference. Rastus, however, had no such qualms, and ambled across, nobody else following; so on reaching the other side, I tied him up, and led the others across in turn.

The second slide, being more overgrown, looked much tamer than the first, and we never even paused. Equine impulses were probably directed towards an extra succulent clump of lupines half-way across - or perhaps the human reaction was that if we didn't go forward, we'd have to go back, which would mean recrossing the *first* slide.

But there was one more contretemps to come. It was Kitty's turn to lead the packhorse, and the rest of us were fifty yards ahead, when we were arrested by a stream of most unladylike language from behind. I was conscience-stricken. The wife of my bosom, the mother of my children - I thought to myself - has sat down in a mud-hole. What a way to start a holiday. Fortunately it wasn't as bad as that - she had only gone in up to her knees, when Gypsy, anxious to get going, gave her a small nudge. English saddle had never involved this sort of adventure.

Fifteen miles from our start-point, the trail meandered upwards away from Gun Creek, through meadows thick with balsam root and arnica - a yellow carpet - and entered the spruce woods, from which the lake took its name; and so to the lodge. The horses did know the way - straight into the corral.

The lodge had seen better days - only half of the roof was intact, the other half lying on the ground where it had landed after what must have been the grand-daddy of a storm. But it was going to be home for the next three days, so we dismounted and stretched our legs.

As for David - no sooner had he unsaddled, than he was out in the somewhat beat-up row-boat which was drawn up on shore; and no sooner was he out in the boat, than he was back again on shore, holding up the first fish.

Now I realize that this is a fishing story, and the very concept of a fishing story suggests that it be taken with a grain of salt. But the plain facts were - and I don't expect to be believed (after all, this happened forty-five years ago) - that inside of ten minutes we had caught our dinner.

Fortunately the half of the cabin that still boasted a roof contained the kitchen; there was a cupboard containing some staples - tea, sugar, salt; and another cupboard containing other staples in various stages of putrefaction (which we buried). But it was obvious that we might as well have left behind our Kraft Dinners, and tins of meat-balls, as we were going to have fish for breakfast, lunch and dinner. We couldn't wait for Kitty to fry up the first batch; and the kids were out of the door soon after, and on the lake again. Another half-hour, and we all had our limits, and enough for a huge breakfast.

We slept soundly in the well-ventilated lodge, unbothered by mosquitos and other nocturnal insects - which, as we found out next morning, had been delighted to find us there.

We had tethered the horses, there being vast areas of grass land around the lake. Some horses get tangled up in the ropes when tethered, but ours were obviously familiar with the idea. However, it is amazing how quickly a horse will eat all the standing grass in a 75 ft. radius; after which he simply stands there looking reproachful. So we had to move them around frequently.

Once on a later trip to the lake, David and I went with a wrangler, who in time-honored fashion staked the mare, and hobbled the other horses, all of whom had bells around their necks. So while David and I fished, Ron hung around the cabin for all of the second day, listening for horse bells; and was able to furnish a commentary every few hours, as to where the horses were. ("They're up at the top of greasy hill - must be heading for the upper Tyax!"). This is all part of the game of horse versus wrangler, to see who can outsmart the other.

However, these horses knew full well that we intended to leave the next morning, so it was not surprising that at 6 am there was not even the faintest whisper of a horse-bell in the air. Ron was equal to the occasion - ("I know where those ding-dong critturs are ... "), and he was off without even a cup of coffee. The four of them were actually not very far away - but simply standing motionless and silent in the timber, a ruse that gained them at least two hours, before they were driven recalcitrant into camp, by a hungry and irate Ron.

The moral is, of course, that the wrangler must be smarter than the horses, otherwise you are afoot.

I have always found fishing to be the most delightful and relaxing occupation. A person does not need to catch fish in order to enjoy it. In most parts of B.C., when one goes fishing, one finds oneself in the most superb mountain scenery in the world; yet the satisfaction goes beyond mere landscape. For instance, on a little lake on the coast called Kokomo, a friend and I had two hours of intense pleasure in the pouring rain, and didn't land a single fish.

A non-fisherman, on hearing such a statement, might conclude that we were both out of our respective trees; and to be truthful, I seldom go fishing in the rain any more. But without analysing the factors that induce fishermen to go fishing, I have to say that sitting in a boat holding a rod, in a high mountain lake in British Columbia, has to be the most relaxing and pleasurable way of spending a day that I know of. Maybe I should amend that statement. "Relaxing" is not quite the right word when one is forever having to pull in ones line with a fish on the end (as in Spruce Lake).

The next afternoon early, there was an invasion. Fifteen teenagers, accompanied by two adults rode into camp. The adults were both doctors, from Gun Lake, and the teenagers all summer residents from the same area. We were glad of the interruption, as so far, a Forest Ranger had been the only other human in the area beside ourselves.

Naturally all had fish for dinner. Kitty and I had in our packsack a small flask of gin (which we brought for medicinal purposes), and we felt that since the other adults were doctors, we could put it to legitimate use, particularly as the doctors were complaining of sore muscles. But while the adults, replete after dinner, could think of nothing better than stretching out, not so the teenagers. The one wrangler, Brian, with seventeen horses to look after, had just about got them hobbled and dispersed, and was starting his dinner, when the word was passed around that horse-racing was to be the order of the evening. I think various people (such as the adults, and certainly Brian) vetoed the idea, but not firmly enough. For the rest of the evening, those who were not fishing (there were only two boats), were busy seeing how fast they could make their horses lope along the trail beside the lake. I sympathized with the horses, and with Brian, who by nightfall was not feeling too amiable.

Next morning, right after breakfast, the cavalcade left. I have a mental photograph, highlighting the varied colors of their anoraks as they rode off along the trail.

Next day it was our turn. Early in the morning we saddled up. On went the diamond hitch, and we made the return trip without incident, and in jig time, slides notwithstanding.

I have been to Spruce Lake on many subsequent occasions, often with David, and using other even more scenic routes over the mountains. In summer, every species of alpine wild-flower is displayed on these slopes. Thus at 5000 ft. elev., in June the meadows are yellow with avalanche lilies and snow buttercups; in July, blue with lupines. Hundreds of varieties of flowers are in profusion, tapering off to the Moss Campions and tiny alpine species as the altitude increases. At 7000 ft. elev. I always marvel at the two inch high pines, which persevere, but will never get taller. Just being in this country gives one a sense of rightness - one doesn't go there simply for fishing.

Some time ago, my friend George and I, with both our sons, were camped by the lake, having hiked in over the high trail, when a small float-plane landed and disgorged a large, pear-shaped man, complete with fishing-rod, fishing boots, fishing hat, landing net, and basket (obviously to put the fish in). But there was something strange about the get-up - for instance there were no flies in his brand-new hat. Moreover, he himself didn't seem too sure about how to use all this brand new equipment. We commented on his lack of camping gear and sleeping bag, but he told us that the plane would be flying him out again in three hours time.

So we were overjoyed when the plane came back later to pick him up, to find that he had caught exactly no fish at all. It was the first time we had heard of anybody being "skunked" at Spruce Lake. But, as George pointed out, it served him darned well right. Everybody who wishes to experience the grandeur of a prospect such as this should have to suffer a bit for the privilege; if one is not prepared to rough it, how can one appreciate its finer aspects?

But we have been fighting a losing battle. No place is remote now, since the advent of the helicopter; the planes fly in and out of this area daily. What of the ecology? The marmots, who observe us curiously from in front of their burrows, and sound their musical warning across the hillside when we pass, do not relish having too many of these ridiculous two-legged creatures in their domain, and many have left; likewise, many deer and other wild-life; and who can blame them. "Over-use" is the key word. Choppers in summer, skidoos in winter. Meanwhile bureaucrats argue - in twelve page booklets - as to the real meaning and proper classification of the concept of "wilderness", while government departments, formed for the purpose of defending the environment, steadfastly avoid taking any protective measures.

A few years ago I remember seeing an advertisement by a small charter airline company, hosting 'champagne lunch on a glacier'. Now I am not by nature a captious person, although I defy anyone to get to my age without ingesting a fair-sized morsel of cynicism; but this ad set me thinking (again) of the bleak future for B.C.'s wilderness. I visualized more and more helicopters landing on mountain peaks, disgorging more and more pear-shaped people; and on an adjacent glacier, an open crevasse with a sign on it - 'Deposit garbage here'.

An exaggerated scenario? It is probably happening already.

HOUSE IN THE WOODS.

"Middle age is when you've met so many people that every new person you meet reminds you of somebody else".

– Ogden Nash (1902–1971)

Lots of people dream of having a house in the woods. We built one; although construction of it took seven years.

It may seem odd in this modern age, when a person wishing to build a home in a growing community only needs the initial down-payment and a willingness to shoulder a mortgage for the next twenty years, that *anyone* would want to build a house the hard way; in an area where mortgages are not obtainable, by paying cash for building materials (as it becomes available), and where labor (apart from ones own) is hard to get; and especially when the building site is three hundred miles from ones usual place of residence.

This "house in the woods" idea must have insinuated itself into my subconsciousness many years ago, as a retirement ambition. I had bought the property in 1956, when access to the Bridge River Valley was via the P.G.E.Rly flat car service from Lillooet to Shalath, thence by the road over Mission Mountain, and alongside the Bridge River (as yet undammed to become Carpenter Lake). At the head of Tyaughton Lake stood Mrs. Keary's lodge, and thither I made my way that summer. She gave me a bed for the night, and the key to the cottage on the next lot, which was for sale, and accessible by foot trail along the lakeshore. As I walked along this trail, I could hear the "plop" of the fish jumping, and see the water shimmering in the sunlight. Who would fail to be persuaded to buy in such circumstances? There was even an old cottage on the land, habitable in summer, and a small landing stage.

Since I have for some thirty years been spreading around the quite erroneous meaning of the word "Tyaughton" lake, as it was transmitted to me in the 1950's, I think I should now state unequivocally that the word does *not* mean "the place where the fish leap". That the fish do leap is incontestable, in fact there are times when I am tempted to take out a frying pan in the boat instead of a fishing rod. But I must set the record straight. "Tyaughton" (Tyax for short) is an anglicized version of the Lillooet Indian word for "the farthest place upstream that the salmon go" (i.e. their spawning ground).*

In the early years, we never even considered building here. There was, in any case, no road access. We had a 13 ft. Boston Whaler type boat (quixotically named "Kwitchabitchin" - an old Skeena River Indian expression), transported on a trailer, and that was access enough when we camped there for long week-ends. But it so happened that in 1965 a bulldozer happened to be travelling back from the Empire Mercury Mine on the way to Kamloops, and stopped by the end of the lake. For what today would be a trifling sum, the operator put in a driveway for us. It was over half-a-mile long, but it was road access.

Now we had a road; and from there events simply escalated. If we had a road, why not some clearing; and if we had a clearing, why not another house; and a wharf; and a shed . . . . . .

Actually, it is doubtful whether we would have embarked on this ambitious project had it not happened that in the '70's our children were poised to leave home, and we - the old folks - were about to experience that void which always occurs in such circumstances. Nature, they say, abhors a void, and so it was that we looked to Tyax to fill it.

Bjorn was a Swedish lad, a thoroughly nice guy, and it is noteworthy that for all the time he helped me, on clearing and burning at Tyax (which we did by hand), we never had a wrong word. For three weeks that winter we power-sawed, chopped brush and built fires; in the evening we played cribbage, or read books or fished till it was dark. Bjorn had a highly illegal system of suspending two lines with hooks and worms from a piece of driftwood, which he then turned loose to drift overnight. In the morning, he would take the boat out and look for the driftwood. Sometimes - if both fish swam in the same direction - it could be half-way down the lake. I have never dared to use this technique since then - maybe it is legal in Sweden (though I doubt it). But I certainly helped to eat the fish, which never failed to get caught, and ended up on the hibachi, with a few spruce needles thrown on top of the coals for flavor.

It is not only icebergs that conceal seven-eighths of their bulk down below. In the middle of our house site was a small rock, no more than twelve inches around at the top. But the moment we brought in a backhoe to dig footings, all hell broke loose. The small rock turned out to be seven feet square by seven feet deep, and in tackling it the backhoe overturned. The owner of the machine (we were not present at the time) then departed in search of assistance (and probably liquid refreshment), and while he was away there occurred a windstorm of mammoth proportions. It went through our weakened grove of trees like a typhoon, blowing at least twenty of them across the road. By the time the bulldozer arrived, the site was looking like the aftermath of Hurricane Hannah.

All this was, of course, just preparation. It would be two years before we actually built anything, Meanwhile, it took four years of fall burning to get rid of the stumps. A curious thing about stumps is that they start by being about four feet across, including roots; and each time they are burned they simply get smaller, ending up as miniature black, club-shaped objects no bigger than a fist.

We set aside a week of summer holidays for the concrete work. Actually the word itself - "concrete" - was enough to suggest problems in those days. We borrowed a small mixer of about a quarter yard capacity, driven by a Briggs & Stratton engine. It worked. The pump, however, which we rented, did not; and my wife always boasted of how she packed all the water for that cement in buckets from the lakeshore. As for the aggregate, a fair amount of it consisted of lava-ash, a local product that had abounded in the valley since Mt. Meagre blew its top several thousand years ago. The other laborers were my son and a friend - both sixteen years old. They did Trojan work. Moreover David by the age of seventeen had quite definite ideas about the construction business (as I heard from another source). He said that he wouldn't mind sharing a house, or renting a house, or buying a house, but on no account, ever, would he *build* a house. It took me five more years to come to the same conclusion.

In 1974 a contractor was not difficult to find. One, whom I knew in the way of business, was more than willing. He had a crew, he said, who were all keen to spend the day working their heads off, and the evening feverishly catching fish. In fact they did work non-stop, and had the building completed in five days. As for the fish, they must have cooperated too, because the crew went away quite happy; leaving us with four walls, a roof, floor, doors and windows in place, and inside - stud walls. The rest was up to us.

Whereas the road had taken one year, the clearing a second, the framing a third, we were to spend five more years on "the rest" before the house was completed.

First came the plumbing. The previous fall I had enrolled in a plumbing course, for two evenings a week, taught by a retired master-plumber, an experienced if rather dour character. I learned enough on the course to plumb a small dwelling, which was all I needed. The proceedings in the class were never very exciting, and I used to try to enliven matters every now and then by interjecting a cheerful , and maybe slightly humorous remark. My intentions were never derogatory, but, looking back, I don't think I was the instructor's favorite pupil. I remember one evening when it had snowed during the day, entering the classroom, and remarking that it was "a nice day for frozen pipes". I have often wondered since if such a subject is one that makes *all* plumbers fighting mad. Certainly it drew a reaction from my instructor, whose look of pure, unadulterated hatred said quite plainly that, had the other pupils not been present, he would have clobbered me with the nearest pipe-wrench.

I had as a helper my son's friend Mike. He also learned about plumbing. He might even have learned some new words, too. But I have to boast that the system we installed only once developed a problem - due to a defective pressure tank, not our fault - and on only one occasion did we had a frozen pipe. Now I know why plumbers are sensitive about this subject.

After the plumbing came the wiring; after the wiring, the insulation; after that the interior finish; and the ceilings; and the floors; and the kitchen cabinets; and the appliances - and everything *including* the kitchen sink. Had we worked on the house to the exclusion of all else, we probably would have finished in a shorter time. But there were many interesting things to see and do in this Tyax country, so we took our time over building. It took us another five years before we could say it was actually finished; and, of course, a house is *never* finished.

As for premises , we ended up with vegetable garden, tool shed, wharf, barbecue, hanging baskets, flagpole; we (and the Queen) would usually pull down our flag when not in residence. We even had a lawn, and in common with other suburbanites, mowed the darned thing every week or so. But in this neck of the woods, we did not do it to keep up with the Jones's. We did it in self-defense, otherwise the jungle would reclaim its own; starting with the fireweed, then the berry bushes, then the willows and finally the jackpines. I figure that twenty years is all it would take for the forest to grow back.

I still wonder at times why we undertook this vast project. Living in the house which one has built can give a sense of supreme accomplishment, and that is very satisfying up to a point. But one can hardly sit back and enjoy the fruits of ones labor for twelve hours a day. Sooner or later one has to ask:- "What next?" Retirement can be absolute, or it can simply be a coffee-break in an unending procession of events. In the latter case (our own view), sitting in the rocking chair has to be the most monumental waste of time. In other words, there must be *another* project.

But where houses are concerned - one is enough.

WHEN TO STOP WORKING.

"A man's homeland is wherever he prospers".

– Aristophanes (450–385 BC).

If, as the sage would have it, life begins at forty, what - in his wisdom - does he prophesy for sixty-five ?

It doesn't take a savant to know the answer to that - this is the era of the middle-aged spread and the aching back, when one loses hair, and teeth; and can contract some of the least desirable ailments in the medical dictionary. So far, all I had had personally was a back problem, which is probably a good ailment when one is sixty and with a tendency to be overly energetic, as it ensures that one slows down. The alternative is to suffer pain, and therefore, theorizing that pain is a warning signal, a bad back is simply an indication that one is overdoing matters.

Nathanial Gubbins (a pen name), who used to have a syndicated column that appeared many years ago in the Vancouver Sun, used to write every so often a "letter from my stomach"; I used to find it very entertaining. With me it is my back, and if my back sends me a message, I know exactly what it means, as we are on the same wavelength physically; as Nathanial was with his stomach. With him, it was usually a request to stop sending down such exotic tidbits as Lobster pate, Black Forest cake, smoked salmon and Cheshire Blue cheese - not necessarily in that order, but accompanied by several Martinis. Not that my back would have a problem coping with this sort of treatment, but when I do too much chain-sawing, or digging potatoes, it usually tells me to stop, and I have learned to take its advice.

What I had started out to say before the digression was, that when one reaches sixty years of age, two notions, which in earlier life were only considered as abstract, become important practically. One is the (physical) need to slow down, the other is the prospect (pleasurable or otherwise) of retiring from the daily grind. In the case of many energetic sixty-year-olds, the latter idea often means exchanging paid work for voluntary work; but it should mean working at, or doing projects which a person wants to do, as opposed to things which he must do, whether he likes them or not.

A work-a-holic is a person who takes work home every evening. I was not one of these, I was a worrier (show me a surveyor who isn't). But city surveying - always a potential source of worry - was becoming legally and technically more and more demanding. Even if we didn't take work home in the evening, it was only too easy to take our worries home. The solution was to sell out, and in the end, at sixty-three I retired from the partnership.

So when my partnership was ended (and in order to avoid the 'empty nest' syndrome), my wife and I moved to the Bridge River country, where we could enjoy the mountain air, the vistas of valleys and peaks, the fishing, the water-skiing, the hiking trips, the horse trips, the cross-country skiing and the life of Reilly in general. So what was missing? That is a hard question to answer. Some, my wife is one of them, say that I am never happy unless I am beating my brains out on some project or other. That is probably close to the truth.

Whatever the reasons, the result was the same. I went back to work. I opened my own office again - one room in the house. The object was to take on one or two jobs in the summer months, enough to keep occupied, and hopefully earn a few bucks to add to my retirement income. Every person has their own peculiar trait. Mine is a rare one. I *missed* being out in the bush.

Working directly over original Crown Grants always interested me, as it meant having to re-establish boundaries set up fifty, sixty or more years ago. I always found the rerunning of an old line, finding evidence of old tree blazes and axe-cuttings on the way, and finally locating the old corner post itself, or the pointed base sunk in the ground, to be one of the most rewarding aspects of land-surveying. Bearing trees, which are blazed at the time of the original survey, and carved with the distance to the corner post (such data being recorded in the official field-notes), had always been most valuable evidence for finding old corners in British Columbia. But now that the logging industry is turning all the forest land in the Province into wide open spaces, it is often an exercise in futility to expect to follow a line by walking to its conclusion (as surveyors have been able to do for a hundred or so years), to where a relatively undisturbed post awaits. Instead we find a logged off wasteland, where even the bearing trees have been cut down. Even so, surveyors of my vintage, who by training and experience are part sleuth, can enjoy such a challenge, and it rarely happens that such a corner, having first been established mathematically, cannot be defined properly by finding physical evidence on the ground as well.

In a place such as the Bridge River valley there is little call for a surveyor's services from purely local interests. On the other hand government agencies are all the time requiring legal surveys in such places. In seven years I kept busy every summer, and did a lot of interesting projects not the least of which was the development of a new lodge at the end of Tyaughton Lake.

It was plain that this valley could not remain an undiscovered beauty-spot for long after the 70's. The Carpenter Lake road had ben partly paved, and the vaunted Hurley Pass had - by a succession of logging operations - been opened to the public. Every resident of the the Bridge River valley welcomed this. We had one of the first Chevrolet Blazers, with a winch in front, and put in many a day exploring new and old routes, safe in the anticipation that we would never get stuck; and we never did, but only because we always carried two spare wheels.

So with better access development was bound to occur, and one of the first to sell out was Joe Bingham, whose holdings included a small lodge on the west side of the lake and a lot of vacant land. There was to be a stupendous log lodge development, picturesque and entirely worthy of its setting . The developer was Swiss. the new owner German, who it must be admitted knew little about running a resort (he soon learnt). The whole scheme seemed to fall into place quite logically. Scott MacKenzie - no mean log builder himself - was construction superintendent. I laid out the footings for the building, and using the techniques of modern log building (precut logs transported to the site), local labour put up the building that now dominates the north end of the lake. I give Erse full marks for planning such a striking development, completely appropriate to the site. I have seen lots of development in my time, and perhaps a person should not speculate on what might have graced that location if it had been badly handled.

Such a project requires a lot of other surveying - a subdivision plan with all the governmental approvals to be acquired. Was I supposed to be retired? If so - I immediately came out of retirement; and what could be more appropriate, as, in this remote area, I lived only about a mile away from the site.

So many were the activities that grew out of the lodge's inception. I remember cutting trail right around our side of the lake, so that parties of horsemen from the lodge could ride the circuit. I also recall with much nostalgia a New Year's Eve party, when Kitty and I put on ski clothes over our party duds, and skied down the icy lake in brilliant moonlight; we entered the lodge by the basement door, left our skis below, and joined the dance On the way back - there - in the middle of the lake was a bonfire, and Scott with his kids were sliding up and down the lakeshore banks, and toasting hot-dogs at the fire. Fortunately we had extra wax, as the wet snow clogged up our skis, but we helped with the hot-dogs that New Years morning.

There are many nostalgic memories of those Tyax days. Every year there was a summer party, and the party-giver invited all residents to it. We found that not only were the other residents interesting people; most were outright characters. For instance, as an example of stalwartness (if there is such a word), Scott and Petie pre-empted 640 acres of Crown Land, seven miles up the valley, as a farm (the only level place - I surveyed it for them), built a log house, kept livestock, raised a family, and are still there. All had one thing in common - a belief that the valley was unique and a desire to live there.

As for the lake level. I understand it is a common state-of-affairs on B.C. lakes to have two factions among the residents, one in favour of low water and one in favour of high. Tyaughton Lake is no exception. In our case I was instructed by the Provincial Game Warden to drill a hole in rock. to mark the arbitrary level which had been agreed to. The hole is still there, but it means little, as, one month after I had drilled it, it was repudiated by Hornal and family as being too high, and sundry others as being too low. All too frequently came the sudden drop in level, after Hornal and his friends got together to destroy the beaver dam. I well remember - after Steve, Bob Sly and others had rolled two enormous boulders into the lake outlet, feeling certain that no-one would be able to tamper with them - a large game warden with handcuffs on his belt came knocking at Steve's door with:'Did you put that obstruction in the spawning stream at the end of the lake ?' Needless to say, we had to remove the boulders, although no-one admitted any wrong-doing.

Several times the Vancouver Natural History Society held their annual alpine summer camp in our area, and on other occasions we (four sixty-five plus year olds) managed to climb Mt Penrose and Mt Truax. Moreover we got to know the surrounding country pretty well, with horse trips to Warner Pass, Mt Sheba, the upper Tyax, the Dill-Dill plateau, Lorna Lake and one longer trip to Dorothy Lake (near Chilco), where we swam the horses across the Taseko River. That, I recollect, was a cold, and of course wet, crossing, and when Barry Menhinnick passed me an open bottle of rum after landing, I eagerly took a swig. It was overproof, and I didn't stop coughing afterwards for the next five minutes.

I have been asked, sometimes facetiously, what I see when looking through a transit telescope. Although ones impulse is to reply in the same vein (the time-honored joke used to be "an election"), what I really saw was British Columbia during forty of its most developing years, an experience I could not have had in any other walk of life, nor would I have willingly traded it.

However, in 1989 (in my seventies), when we said goodbye to Tyax for what I hoped would be our last move, I decided to sell my transit and other equipment. I hated to do it, but there comes a time . . . . . .

Now, looking back from 2012 and in my nineties, with activities still crowding in, there is still much to write about, and I sometimes wonder how in those days I managed to find time for working.

GENEALOGY.

"You cannot choose your ancestors, but then they probably wouldn't have chosen you".

– Anon.

I think that most people have more than a passing interest in their forbears. For a historically minded person that interest would be greater. My own interest went a long way back, as our family were in possession of a 'tree' which purports to extend back to the fourteenth century.

For the uninitiated a family tree is in the form of a tree cut off at the stump, and inverted; it has the patriarchs at the top, and new arrivals at the ends of the downward-pointing limbs. The person for whom the tree was commissioned is, of course, at the bottom centre, and the tree is unique to him, as his sisters, cousins and aunts all hang to one side or other. Thus, for instance, my tree will have me at the centre and my cousin to the side, while his tree will have him central and vice-versa, although other branches will not differ. The person at the bottom of the Cotton family tree is Cecil Cotton, my father's cousin, unmarried so unfortunately without issue.

Cecil died in 1954. He was a very competent and likeable business man, who knew what poverty was in his younger days. He gave me my first job after leaving school, for which I was most grateful, as I am sure no-one else would have hired such an aimless character as I was then. I had great admiration for him, and for the history which he unearthed, and passed on to us as family lore. For many years he was an executive member of the Genealogy Society of London, and such research was his lifetime hobby. The tree which he commissioned is scrupulously detailed, and is incredibly interesting.

Basically, the name 'Cotton' is derived from the ancient Saxon word for a small house, and there was a Coat-of-Arms, a crest and a motto. All are shown on this family tree, which is in reality an historical record of a family in Co Cambridge, England. It includes records of one William Cotton, Vice-Chamberlain to King Henry VI, who died in the First battle of St Albans, a priest, a priest and martyr, sundry members of parliament and county sheriffs, an Admiral of the Fleet, and a physician whose patients included Royalty. There was even a Cottonian Library (now in the British Museum; and there were two baronetcies, both of which died out.

Here, again should be explained that a baronet warranted a 'sir' before his name, and such a privilege was handed down to the eldest son. But a baronetcy was not given by the monarch as a mark of honour. It was bought (James I initiated the idea in 1611 because he was short of cash), so the prestige was purchased rather than earned.

Another interesting fact also emerged from a study of this family tree; there was no doubt that the Cottons were particularly adept at acquiring estates by marrying the only heiress. In this way they acquired Landwade, and later Madingly - that huge estate that included a building as big as Buckingham Palace.

This latter estate was latterly held by Admiral Sir Charles Cotton. A person should be proud to have such an ancestor, and I was (although he was quite a long way off to the side of Cecul's family tree). He joined the British navy in 1772 at nineteen years of age as a midshipman, and in those brawny days of frigates and real sailing ships, he rose to the highest rank, and had a very distinguished career. Cotton Point on Keats Island is named after him, although he did not voyage to British Columbia, and a detailed description of his exploits is given in Walbran's B.C.Coast Names. Remember 'Mutiny on the Bounty'? Admiral Cotton sat on the panel at Captain Bligh's subsequent court-martial. His wife was Philadelphia Cotton, and his eldest son (Sir St Vincent Cotton - 'Uncle Vinney') - inherited the title. The latter was the subject of many a tall story; he is largely credited with gambling away the entire estate; after which his mother and sister bought the estate back, and enabled him to do it again; although it seems that finances had progressed downwards for some time before Vinney really got going. He died a confirmed profligate, and that branch of the tree dies with him.

Of course, an illustrious ancestor does not necessarily pass on his talents to his descendants. Often the opposite result occurs (as in the case of the Admiral and his wastrel son!). At a younger age I spent some time studying this family tree that Cecil Cotton had provided - at arms length, as it were - as there was no way that I could really associate these shadowy figures with my own part of the family that I knew. But it was interesting to know about these vanished Cottons, even if their estates, holdings, pomp and circumstance had already faded out.

In January of 1975, Kitty and I were in England staying with her brother, and as Landwade and Madingly - the old estates of the Cottons was in the vicinity, we resolved to pay a visit.

The colossal building at Madingly now forms part of Cambridge University. We found our way there, and after knocking at the door, and explaining who we were, we were greeted by the Warden's wife who called out over her shoulder:-

'Henry ! Come on down - I've got a real, live Cotton here !

They gave us the grand tour; and a cup of tea.

After that we visited Lanwade, where there was no building. only an empty lot. Local lore remembers that many years ago the building was razed, and the creditors were waiting on the site to claim the building materials. Again we were entertained by the incumbents, and then shown over the church.

Now this little church is really worth a visit. It is Saxon, built in the 12th century, and stuffed full of historical records, as well as relics of the original Cottons of the fourteenth century. It was the high point of our search for 'roots', and although I have never been keen on looking inside churches, we spent two timeless hours there.

Genealogical research has become so much easier in this day and age, with the advent of computers. The 'tree' that Cecil put together was validated by an enormous amount of hard searching in old parish records, family papers, and official documents. I have to mention here Ross Cotton, who is a genealogist that only deals with the name of Cotton. At first I thought that this would have to be a fairly restricted field, but it turns out that there are more than 30,000 Cottons in the USA, and corresponding numbers in Canada and Australia. So today's research consists in large part of comparing DNA's to find out who is related to whom. I had some correspondence with Ross in 2005, but if there's one thing I would like to keep to myself its my DNA. So I did not part with it.

Now while I agree that DNA comparisons will group existing Cottons together logically, I do not see how it connects with the past; for instance - how do you obtain the DNA of someone who has been dead and buried for 200 years ? So it does seem that Cecil's 'tree' is still valuable evidence, as well as being an interesting asset, and we should regard it as such.

BRITAIN IN RETROSPECT.

"Contrary to popular belief, English women do not wear tweed nightgowns."

– Hermione Gingold (1897–1987)

When I lived in Britain, I accepted the country, its institutions and social structure, without question. It was my birthplace, and I thought very little about it, one way or another. But after a long absence, and particularly when one has ceased to look upon the place as "home", ways of living which one formerly took for granted, come to be regarded from a different standpoint.

Now, while many of the British institutions are praiseworthy to say the least (as many eminent writers are apt to point out), I am ashamed to say that I have never returned from a visit to the old country without some or other cause for irritation; and lately I have found that the sensation of sinking into a seat on the return Air-Canada flight at Heathrow, and being exhorted in Canadianese to make sure that my seat-belt is securely fastened, is every bit as pleasant as any highlights of the trip that is just over.

Our first visit to the 'old country' was at Christmas in 1956, ten years after I had left it as an ex-serviceman with few prospects, and only enthusiasm as a substitute for experience. During the ten years I had become relatively successful in my profession. Furthermore, my wife and I now were living in our own home, and were the parents of two happy and healthy children. What better idea than to visit our "folks", who would be so happy to see us, and rejoice in our achievements? Kill the fatted calf - the prodigal returns. We would really wow them!

Of course, it did not quite work out that way. Had there been a fatted calf, my father would probably have given us a piece off the rump. But circumstances were not that easy. My father with his second wife, lived in the south of England, as did my mother (relationships being somewhat strained between them); while my wife's brothers, and their wives, lived in Hull and Harrogate respectively. We were to have three days at each location, and a lot of travel by rented car in between.

My father did his best. He fixed up the frigid, spare bedroom upstairs for us, with two cots for the children aged three and one-and-a half respectively. But how can one relate to a person whose principles included the one about children being "seen and not heard", and whose bête noir was anyone with an "American" accent?

At my fathers house, we found it quite hard to relax. David expressed his indignation early the first day, at breakfast, by throwing up down the front of Kittie's dress, and inside at that. My father bore up bravely, a pained expression on his face. Kitty left to change her dress and David, amid a well-bred silence. I comforted Wendy, who was showing signs of wanting to cry. I felt like doing so myself, in fact. However, on the second day of our visit, I resolved to tell my father that however repugnant the role might be to him, he was my kids grandfather; and did so. Which remark regretfully caused quite a rift in our association, so that we hardly corresponded afterwards; and I never did see him again before his death in 1974.

And on the way south via the A.1. highway, with the windshield wipers ploughing the smog into twin borders of black soot, it was my daughter's turn to throw up - all over Kitty's nice fur coat. I think the incidents that remain in my memory the most, on that inauspicious visit, are the throw-ups. David had started the ball rolling before he even got to Heathrow, on the plane from Amsterdam, and again Kitty absorbed most of it. It seemed that the kids were not very impressed with jolly old England; and I must record, nor was I. When the visit was over, I resolved never to darken its shores again.

For some twenty-five years, I kept to this resolve, although we did keep in touch, and had my mother come out to Canada for some five trips in the summer. Then finally in the spring of 1975, my wife and I went there again. I don't know if Britain had changed in the meantime, or whether it was myself. I do know that a week on my cousin Maureen's farm at Bryn Teg, with as rowdy and natural family as any I have known, convinced me that Wales (if not England) was still a healthy place, and contained people to whom one could relate.

Since then Kitty and I went regularly to Britain at intervals of three or four years. We both saw our relatives. My mother passed away at 100, in the Isle of Man, getting two citations - one from the Queen, another from the Governor of the Isle of Man.

We always went to Wales. We travelled on the London Underground, saw the Festival Hall, the "Bloody Tower", the whispering gallery in St. Paul's Cathedral, the Cheshire Cheese, the Horse Guards Parade and Buckingham Palace. We marvelled at the Cotton family seats at Landwade and Madingley; ate porridge and kippers for breakfast in Dumfries; walked Yorkshire moors; visited the grave of David of the White Rock, and absorbed countless "ploughman's lunches", not to mention pints of ale and "whiskey-macs".

I am now too old to go again. Moreover, I have now no relations except second cousins - Maureen's daughters Non and Mair (with whom I correspond). But what was it that drew us back when we regularly visited?

On the one hand, there was the hospitality of Wales, with its singing, and the mountains of Snowdonia which I knew as a boy; and the pubs - surely unique throughout the rest of the world; and the beer; and the good fellowship that goes with both; oh yes, and when one goes out to dinner there is that desert trolley that they push around after the main course - that is worth a couple of points, too.

On the other hand, however, there are the teeming mobs of people; and of cars (all on the wrong side of the road); and Heathrow; and the patronizing maitre-d's in restaurants (what has Simpson's of the Strand got that any good Vancouver restaurant hasn't got?); the high prices; the cold, damp houses in which the central heating never works; oh, and those peculiar, clipped accents.

Perhaps the pros are outweighed by the cons. But the odds are unimportant. The fact is that I have been a Canadian for some 65 years, living in the best part of the world - British Columbia, where I have brought up a fine family, who have in turn brought up their own families too. So – Britain, for me, is just another tourist mecca; moreover, the store of knowledge that I possess about the 'old' country is now largely inappropriate as it pertains only to the past.

GULF ISLAND

"...a right, tight little island"

– Thomas Dibdin (1771–1841)

Looking for property on Saltspring was easy. One could phone an island real-estate agent, catch a ferry as a foot-passenger, and spend the day being driven around, looking and generally becoming acquainted with the island.

We settled for a house on a hill, overlooking Trincomali Channel. The price was reasonable, because it had a shared well with one of the neighbours. Owners with shared wells have been known to come to blows, we had learned, so we decided that if we liked the house (which we did) we would drill our own well. To this end, before purchasing, we contacted the local witcher, Albert Kay, who came round complete with willow sticks, and found s spot near the house where, he said, we would get water at 125 feet down. We put in a peg.

We certainly made the right decision in hiring Albert. After we had bought the lot, and brought in the drill-rig, we hit water at 126 feet down, for 3 gallons a minute - a quite adequate amount. Albert professed not to remember anything about witching this spot for us, and we found out later that our neighbours on one side had gone down 450 feet and got salt water. So - were we just lucky, or was Albert just an infallible water witcher ? We may never know. Anyway - we had water.

Since this was to be our second retirement home, we resolved to make it comfortable. We had half-an-acre. So I put in 2 apple trees, and 2 plum trees, Also various current bushes and plants we brought down from Tyax, and proceeded to make a nice garden; then added a car-port and work shop, and put extra siding on what was a fairly old house.

After which, we thought it was time to find something else to do besides work. We joined the Trail and Nature Club, the Tennis Club, the Historical Association and the Saltspring Singers. Soon we would get ourselves a 25 ft power boat with a deisel, and join the Yacht Club, too; where Wendy and Kevin had been members for many years.

Then the Arion Male Voice Choir came from Victoria for a visit. Now in Vancouver days I had been a long time member of the Vancouver Welsh Men's choir, leaving it only because we had moved to Tyax; and I still felt that real choral singing consisted of the sound of male voices. The Arion sang in the United Church, and after the concert was over there were three of us - Saltspringers - asking to join. There was going to be a problem in attending the Monday night practice in Victoria, as there was no late ferry. However, Harold Traloor, the director, solved that problem summarily by asking for a show of hands as to who would provide the newcomers with a bed on the Monday night.

So each Monday afternoon the three of us took the ferry, dined in Sidney attended the practice and spent the night at the house of one of the hospitable choir members. Kitty and I got to know a lot of good friends in this way.

Piano-playing also came to the fore at this time. Formerly as a member of the Alley Cats band, I had played (and arranged music) in North Vancouver. We had a violinist (a real fiddler), a sax player (who also sang), a clarinet player, a man who played the spoons and a large lady who played a large double-bass; a congenial group and we did have fun, playing at 'seniors residences' each month, and practising regularly. So – when Kitty and I moved to Saltspring, I looked for a similar group to join. But I didn't seem to find one; until I found that (in Victoria) the Oak Bay Seniors had a band, and were looking for a pianist of my calibre. They practised on a Tuesday morning, so soon I was extending my Monday night's stay to take in another practice - arriving home on Tuesday afternoon. It seemed that I was not able to relax as a retired person should; but Kitty was finding her own activities on the island, and didn't seem to mind.

Here I shall digress in order to clarify for myself the purpose of these memoirs. Originally it had seemed to me that since my life was including so many unique experiences, it would be a pity not to record them for posterity, and I hoped in writing to make the various episodes sound interesting. This entails a great deal of typing out the letter 'I'. But no person is a single entity, and as I continue with this series, I realize that what I now write about consists mostly of family matters. However, to alter the context of what I record completely is not possible at this stage. Kitty has been gone for 5 years, and my family - our pride and joy - have long been grown up. So I continue with the 'I's', much as I am tired of seeing them.

Kitty's ideas on family matters were very sound - she believed that once kids have grown up, they must be left to live out their lives without interference, and parents should only offer advice if asked. The system worked very well in our case. However, we do need to see them from time to time, and that was one reason for moving from 'the boonies' of Tyax to Wendy, Kevin and Korena's Saltspring.

Meanwhile, the world had progressed and would have passed us by, but one day Wendy said to me: "Dad, you have got to get a computer!" Till then, though I had been familiar with electronic calculators, computers were a closed book to me. She got me one of the first "Macs" (now a classic), and I found it (a) unbelievably useful and (b) completely frustrating - (that second attribute still occurs time to time). But whereas in the old days I typed out records, and kept them in a file, now I could do everything under one umbrella. That, in fact, is what launched these memoirs - and some more writing.

I had previously written a few articles for B.C.History. The first was about Lajoie, the con-man who has a lake and dam in the Bridge River named after him (in Tyax that story had always interested me). I stepped up the output, and contributed about a dozen articles to this magazine, mostly about old-time surveyors. Then I self-published these Memoirs - entitled Beating about the Bush - using Kevin's publishing designation 'Cranberry Eclectics'. I rented a laser printer, and printed enough pages both sides for 25 books, which Wendy and I put together with curlicue bindings. After distributing them to family and friends, the feedback was very positive, so I took the book to a printer, and had 70 more copies properly bound; and these I sold at the next BCLS convention for $20 each recouping my expenses.

Now, I had always realized how unique was the Hermon firm, as they were the first surveyors to lay out Vancouver City. It seemed a pity that a story so well rooted in history should be allowed to die out. Using the voluminous offices records (we had files going back to 1895, the birth of the City), City and B.C. archives, I was able to put together a historically correct, and authoritative book on the origins of Vancouver, together with maps, photos and illustrations - First in the Field. Many books have been written on the subject, but this one did have a different perspective as it concerned the actual layout of the city. I self published 100 copies of this book - which took a lot of time and effort to research - sold most at the next convention, and got rid of the rest simply by circularizing libraries in the smaller towns of the Province.

But this apparently was not enough for me. John Whittaker had been chairman of the Historical Committee of the B.C.Land Surveyors Corporation, and had just published biographies of the old surveyors active from 1895 onwards. Before he stepped down, the committee made a resolution to work on a second book, that of the first surveyors in B.C. - the 'L.S.Group', active between 1850 and 1895. Various committee members volunteered to research and write the biographies. I was one.

After John stepped down, Bob Allen took over as chairman. I don't quite know what happened after that, but it seemed that the other writers seemed to drop out of the running, one by one, leaving only yours truly. However, I stayed with it, partly because I was really profoundly interested in these stalwart characters who laid out this wild, untamed land in such an orderly fashion, and partly because I was obviously just, plain stubborn. I ended up doing 75 % of the researching and writing, including getting photos and illustrations. I always got the idea that the BCLS Group were not really interested in the publication, but after all, the resolution had been made, and personally I considered that we surveyors ought to know something about our predecessors. It did take a long time - about five years - but thanks to Dave Morton, who was the editor, and sorted out the material which I sent, it got done.

All of this, of course, added another dimension to my Victoria caper. In the end I was leaving Saltspring on Monday morning to visit the BC Archives, and spending altogether too much time in the basement at home, computerizing.

So when we bought the boat (an Albin 25 with a simple diesel engine, and a speed of 7 knots maximum), we needed to get away from the books. For two years we spent cruising the Southern Gulf Islands, Texada Island - Jervis Inlet - Newcastle Island and over to West Vancouver to see David who was now back from the East and living there.

Sometime after this, Kitty became ill. The stroke she suffered had consequences, and she started losing her sight. We bought a chairlift to access the basement and I took over the running of the house, but what was needed was a change of venue.

Meadowbrook was a newly built seniors residence, conveniently located in Saltspring, and they were looking for residents. We put our names down, and in due course put our house up for sale, invested the proceeds and with sundry garage sales proceeded to divest ourselves of items which had taken a lifetime of choosing to accumulate. A sad course, but the only one.

At Meadowbrook there was a meadow (across the parking lot) but no brook. It offered chore-free accommodation. but there was no denying its real function - it was an Old Folks Home; complete with quiet restrained atmosphere, uninteresting meals and a quota of 75% little old ladies. Actually we made several good friends here; and I did try to initiate some stimulating activities - a weekly musical video show, a sing-a-long and a 'happy hour' (although not too many felt comfortable with a glass of wine, and some actually spurned it); and I played piano in the lounge before dinner several times a week. We had social events, and visitors, and family kept close to us.

The home-care nurses came and went all the time. Kitty was well looked after, but she went downhill in the two to three years we were there, and on an early morning in June 2006 my wife of 57 years passed away. She had been a staunch friend and supporter to me and the family for all that time. The memorial organized by our family, and attended by friends from far and wide, did show our appreciation for such a wife and mother.

I did not stay long in Meadowbrook after that.

THAT'S IT !

*"But all he had was that he was a philosopher,
yet had he but little gold in coffer."*

– *Chaucer. (14th Century)*

All memoirs should end on a philosophical note, and what more such remark could I make then to reveal here what I now know as fact; that it is possible for two elderly people (in their 80s and older) to fall in love. It is a state of being not often accorded to the aged, and it sets out reasons for continuing to be active. so before continuing with these scribblings I should pause to acknowledge how lucky Daphne and I are in this new phase of our existence.

It started with a visit to Kimberly to attend the B.C.Historical Society's convention; which led to a visit to a cabin on the Gun Creek Road; which led to a bang-up wedding at that same cabin, attended by family and friends from years back. Since when events have unfolded so normally that I can hardly lay claim to being the rolling stone that I profess to be.

Now, will what follows continue to be philosophy, or will it be simply the ramblings of a ninety-year old ? This has been an Odyssey – the random memoirs of a rolling stone, and it must come to an end. I am more fortunate than Odysseus, who had all those suitors to contend with when he returned home, as my first wife accompanied me on most of my travels. But this is no Homeric epic either, only an account of a life; a life in the past, however, and in that context there are significant differences, worth enlarging upon here.

A generation ago, every-day living held many differences from today. Were we more relaxed ? I do believe that we were. Certainly life was slower, and it seems to me that both activities connected with earning a living, and social enterprise were simpler and more understandable. Although science had made inroads on the chores of daily living, labour-saving devices were few and far between; we pushed no buttons - there were none to push. We were, in fact, on the threshold of todays modern world, and our age-group experienced a lifetime devoted to change.

In my young days, a person did not include milk in his or her shopping list, as the milkman with his well-trained horse stopped at the door, delivered your quota, and took away the empties. I mention this for nostalgic reasons, and because it gives an impression of a leisurely existence. Again - was it leisurely ? We also had the breadman, the iceman (before refrigeration), the coalman and, quite often a green-grocer's cart would call - a quite illegal operation today. We still do have the postman - who is becoming more and more redundant every day - and more garbagemen (who may well be millionaires) than we ever had before. Perhaps matters were not so leisurely after all, as this was the beginning of consumerism, as practised by the merchants themselves (in the absence of planned townships developed around automobiles and shopping centres).

Cities were indeed a lot smaller. London, England, contained about two million people, and was considered the biggest city in the world, although New York was obviously taller. If there was a shopping centre in London, I never saw it. Shops sold their own merchandise, the butcher, the grocer, the baker, the fishmonger, the green-grocer, but they were usually clustered together on the main street of town. As for cars - they were for the well-to-do, and few had self-starters; you turned a crank handle in front of the bonnet to start the engine, and you could end up with a broken wrist if the engine back-fired during the process; and when driving you used hand-signals out of the window - turning signal lights waited till the mid-forties to be invented.

Radio ? Television ? The former was in its infancy when I was young. Britain's first commercial radio station was Daventry ('2-LO calling'), and you listened in (with earphones) on a crystal set - I often wondered what a 'crystal' was - to sounds that were just recognizable as music. As regards the latter (television) - I don't think there actually was such a word. but the idea would be considered in the realm of fantasy, akin to black magic, a medium that was still believed in by some. As for the idea of a person landing on the moon, a good comedian could bring down the house with that one.

I remember coal-gas as a source of heat and light - even street lamps. I went to bed upstairs with a candle in a candle-stick, and a box of matches, after first being solemnly instructed that I was never to 'play' with the matches; and I felt duly proud of the fact that grown-ups would consider me so trustworthy. I do not remember seeing light switches, and certainly not power outlets, till I was in my teens. Telephone – yes, we had it, and paid for each and every call, so a person learnt to be frugal in his conversations (I still am). Dial-up phones eventually came about; but before that, when you picked up the handset, a well-modulated voice said: 'Number please?', and, if not too long a moment went by, might even pass the time of day - a practice few people would have time for nowadays.

My generation were the baby-boomers of the First War. We were brought up in depression and/or poverty, and we went through a war (the Second); but after that we were fortunate in being able to make a good life in a growing economy. The baby-boomers of the Second War may not so favoured, as they grew up in a land of plenty, and now that they are about to retire, the economy has downturned.

It is tempting to say that at my age I have seen it all. But no man has ever seen it all, only parts of it, and often disconnected parts at that. The older a person gets, the more he feels he should 'put the world to rights', but though I do my share of ranting at parties, I leave it to the newspapers to put those ideas into print. However, a little philosophizing is what I had intended to do when I started this section, so what follows is relevant.

The word philosophy is from the Greek, and it means 'love of wisdom'; I use it in the sense of personal philosophy – a set of beliefs and precepts which every thinking person relies on when dealing with every-day affairs. Here I should note that, although earlier in this book I decried the idea of using the classics in early stages of education, it seems that in my own case some good may have come from it; because, when it comes to personal philosophy, I firmly believe that the Greeks have already said all that was necessary.

Consider the axiom:- "Everything in moderation: nothing to excess.". It is a precept that as far as I can remember I have always applied, to work, social life and every day affairs. Whether subconsciously or otherwise, it has always worked, and led to satisfaction. But the picture is bigger; and at the risk of 'putting the world to rights' - which I said I would not do - it is time for mankind to apply this principle to his way of life, and smarten up while there is still time. For thousands of years, past generations practised conservation, though ignorant of the extent of the world's resources; but more recently, now that we are well aware how finite are these resources, we are hell-bent on using them all up as soon as possible.

It would be nice to end this book on a note of optimism, but like the Greeks, I believe in being realistic. I am sure that Zeus (or whatever name he goes by, up there) must be really scratching his head at the antics of these peculiar animals whom he endowed with computer brains in this 21st century. After all, he gave us perfectly good rules for living in 400 B.C.

Too bad that we have not been listening.

THE END.

Tying the knot

Daphne and I

The Well-Wishers

Kevin and Wendy

Korena and David

Daniel

Gabrielle and Denise

Paul

LITERARY ACKNOWLEDGEMENTS.

The Green Beret - by Hilary St George.

The History of the Honorable Artillery Company - by Raikes.

The History of the Ist Bn Loyal Regt in Italy.

Disaster at Bari - Glenn B. Infield. MacMillan 1971.

"Delina C .Noel, an Appreciation" - by Franc Joubin, in The Western Miner & Oil Review, Aug 1958.

The Accidental Airline - by Harbour Publishing 1988.

Hermon, Bunbury & Oke - old records.

1001 B.C. Place Names - by G.P.V. & Helen B Akrigg.

Discovery Press, 1973.

CPSIA information can be obtained at www.ICGtesting.com
Printed in the USA
LVOW040452211212

312650LV00001B/5/P